About the Author

Karl Seelig is a trailblazer in innovation and interdisciplinary problem-solving, with a career spanning over two decades marked by groundbreaking achievements in technology, business development, and intellectual property. Known for his visionary approach, Karl is the mastermind behind the pioneering $5.8 billion global ringback tone replacement technology, a testament to his ability to turn bold ideas into transformative realities.

Karl's journey reflects a unique blend of scientific curiosity, entrepreneurial acumen, and strategic insight. With a foundation in finance, marketing, and biochemistry, he has consistently pushed the boundaries of conventional thinking, bridging diverse fields like telecom, AI, blockchain, biotech, and theoretical physics. His deep understanding of complex systems and his relentless pursuit of innovation have earned him accolades from prestigious institutions, including recognition from Harvard University for his revolutionary approach to patent litigation and cost management.

As a seasoned executive, Karl has spearheaded high-stakes negotiations with Fortune 100 companies, managed venture funds, and played a pivotal role in mergers, acquisitions, and joint ventures across multiple industries. His entrepreneurial spirit shines in his ability to create sustainable business models while fostering innovation and collaboration. From organizing over 500 panels at global events like Digital Davos to championing ESG compliance in ethical industries, Karl has consistently demonstrated leadership that inspires progress and prosperity.

Karl's foray into theoretical physics with the Twilight Dimension Model showcases his ability to tackle some of the most profound questions about our universe. By merging scientific rigor with creative thinking, he offers a fresh perspective on the interplay between quantum mechanics, general relativity, and cosmology. His work exemplifies a commitment to exploring the frontiers of human knowledge and applying insights across disciplines to solve the world's most intricate challenges.

In every endeavor, Karl Seelig stands as a testament to the power of visionary thinking, unyielding determination, and a deep belief in the transformative potential of ideas. His ability to synthesize diverse expertise into groundbreaking achievements makes him not just an innovator, but a thought leader shaping the future of technology, science, and beyond.

Table of Contents

CHAPTER 1: INTRODUCTION TO THE TWILIGHT DIMENSION MODEL — 6
- The Need for a Unifying Framework — 6
- Existing Theories and Their Explanations — 7
- Establishing TDM as a Novel Approach — 9

CHAPTER 2: STRUCTURE AND DYNAMICS OF THE TWILIGHT DIMENSION — 10
- The Twilight Dimension — 12
- The Reality Dimension — 12
- Mechanism of Interaction — 12
- Emergence of Time and Space — 12
- Quantum Behavior and Turbulence — 12
- Cosmological Implications — 13
- Philosophical and Experimental Potential — 13
- The Twilight Dimension: A Realm of Possibilities — 13
- Energy Flow Between the Twilight Dimension and the Reality Dimension — 15
- Energy Flow and Turbulence in the Twilight Dimension Model — 17
- Energy Flow Across the Nested Spheres — 17
- Tesla Valve Analogy — 19
- Key Mechanisms of Flow and Turbulence — 19
- The Role of Feedback in the Flow — 20
- Interdimensional Turbulence: A Bridge Between Chaos and Order — 20

CHAPTER 3: EMERGENT PROPERTIES IN THE TWILIGHT DIMENSION MODEL — 20
- Emergence of Fundamental Properties in the Twilight Dimension Model — 21
- Interconnected Emergence — 22
- Classical Physics vs. TDM — 23
- Quantum Mechanics vs. TDM — 24
- General Relativity vs. TDM — 25
- Integration of Disparate Phenomena — 25

CHAPTER 4: MECHANISMS OF STATE ACTIVATION — 26
- Mechanisms of State Activation in the Twilight Dimension Model — 26
- Environmental Conditions — 28
- Interplay of Factors — 29
- The Tesla Valve as a Conceptual Metaphor for Resistance in State Activation — 30
- The Tesla Valve and Unidirectional Time Flow — 30
- Nested Spheres and Turbulence as Resistance — 30
- Integrating the Tesla Valve with Nested Spheres — 31
- Implications for State Activation — 31
- The Relationship Between Activation and Observer Influence in the Twilight Dimension Model — 32
- Observer Influence as Energy Flow Modulation — 32
- Shaping Reality Through Observation — 32
- Observer Influence in Macroscopic Phenomena — 33
- Experimental Implications of Observer Influence — 33

INTERDEPENDENCE OF OBSERVATION AND REALITY	33

CHAPTER 5: TIME AS AN EMERGENT PROPERTY — 34

TIME IN THE TWILIGHT DIMENSION MODEL (TDM): A BYPRODUCT OF STATE ACTIVATION	34
SEQUENTIAL STATE ACTIVATION: THE BUILDING BLOCKS OF TIME	35
INTERDIMENSIONAL ENERGY FLOW AND THE DIRECTION OF TIME	35
ENVIRONMENTAL FACTORS INFLUENCING THE FLOW OF TIME	35
THE EMERGENCE OF TIME VS. TIMELESSNESS IN THE TWILIGHT DIMENSION	36
IMPLICATIONS FOR TIME PERCEPTION AND REALITY	36
RESISTANCE IN ACTIVATION AND TIME DILATION IN THE TWILIGHT DIMENSION MODEL	36
GRAVITATIONAL EFFECTS AND TIME DILATION	37
VELOCITY AND TIME DILATION	37
COMPARING GRAVITATIONAL AND VELOCITY-BASED RESISTANCE	38
IMPLICATIONS OF RESISTANCE-BASED TIME DILATION	38
GRAVITATIONAL EFFECTS AND TIME DILATION	38
VELOCITY AND TIME DILATION	39
COMPARING GRAVITATIONAL AND VELOCITY-BASED RESISTANCE	39
IMPLICATIONS OF RESISTANCE-BASED TIME DILATION	40
TIME AND CAUSALITY IN THE TWILIGHT DIMENSION MODEL	40
STATE ACTIVATION AND THE SEQUENCE OF EVENTS	40
THE ROLE OF ENERGY FLOW IN CAUSALITY	41
CAUSALITY ACROSS SCALES: FROM QUANTUM TO COSMIC	41
CAUSALITY AND THE EMERGENCE OF TIME	41
TDM'S INSIGHTS INTO CAUSALITY	42
EXPERIMENTAL EVIDENCE SUPPORTING TDM'S CONCEPT OF TIME	43
DELAYED-CHOICE AND QUANTUM ERASER EXPERIMENTS	43
TIME DILATION IN GENERAL RELATIVITY	43
BENDING OF LIGHT BY GRAVITY (GRAVITATIONAL LENSING)	44
EXPANSION OF THE UNIVERSE AND DARK ENERGY	44
COSMIC MICROWAVE BACKGROUND (CMB) AND INITIAL CONDITIONS OF THE UNIVERSE	45

CHAPTER 6: WAVE-PARTICLE DUALITY AND QUANTUM PHENOMENA — 45

TRADITIONAL VIEW OF WAVE-PARTICLE DUALITY	46
TDM'S EXPLANATION: INTERDIMENSIONAL TURBULENCE	46
WAVE BEHAVIOR: INTERACTION OF POTENTIAL STATES	46
PARTICLE BEHAVIOR: RESOLUTION OF TURBULENCE	46
ROLE OF THE OBSERVER	47
TDM'S UNIFIED VIEW OF QUANTUM PHENOMENA	47
EXPERIMENTAL SUPPORT	48
THE DOUBLE-SLIT EXPERIMENT IN THE TWILIGHT DIMENSION MODEL	48
TRADITIONAL INTERPRETATION OF THE DOUBLE-SLIT EXPERIMENT	49
TDM'S EXPLANATION: INTERACTION OF POTENTIAL STATES	49
MECHANICS OF INTERDIMENSIONAL TURBULENCE	49
EXPERIMENTAL OBSERVATIONS EXPLAINED BY TDM	50
IMPLICATIONS FOR QUANTUM MECHANICS	50
WEAK MEASUREMENTS, PARTIAL COLLAPSE, AND THE ROLE OF OBSERVATION IN THE TWILIGHT DIMENSION MODEL	51
THE ROLE OF OBSERVATION IN ALIGNING ENERGY FLOWS	52

FEEDBACK TO THE TWILIGHT DIMENSION:	**52**
OBSERVATION AS AN ENERGY FLOW MODULATOR	**54**
MAGNETIC FIELD, TEMPERATURE, AND INTERFERENCE PATTERNS IN TDM	**54**
PROPOSED NEW EXPERIMENTS FOR TDM	**55**
CHAPTER 7: LARGE-SCALE APPLICATIONS: COSMOLOGY AND DARK ENERGY	**56**
DARK ENERGY AS CONTINUOUS STATE ACTIVATION	**58**
ENERGY INJECTION AND THE EXPANSION OF SPACE	**58**
EVIDENCE SUPPORTING TDM'S VIEW OF DARK ENERGY	**59**
INTERDIMENSIONAL TURBULENCE AND LARGE-SCALE STRUCTURES	**59**
UNIFYING QUANTUM AND COSMOLOGICAL FRAMEWORKS	**60**
VOIDS AS LOW-RESISTANCE ZONES	**60**
FILAMENTS AS HIGH-RESISTANCE ZONES	**61**
SUPPORTING OBSERVATIONS AND PREDICTIONS	**61**
IMPLICATIONS FOR COSMOLOGY	**62**
TDM AS A COHESIVE FRAMEWORK FOR INTERPRETING COSMOLOGY	**62**
ACCELERATED EXPANSION AND DARK ENERGY	63
LARGE-SCALE STRUCTURE: VOIDS AND FILAMENTS	63
BRIDGING SCALES: A UNIFIED FRAMEWORK	63
PATHWAYS FOR FURTHER RESEARCH IN COSMOLOGY USING TDM	**64**
INVESTIGATING COSMIC VOIDS AND FILAMENTS	64
DARK ENERGY AS EMERGENT ENERGY FLOW	64
TESTING TURBULENCE WITH QUANTUM AND COSMOLOGICAL LINKS	64
OBSERVING FEEDBACK MECHANISMS	64
IMPLICATIONS FOR INFLATIONARY MODELS	64
CONCLUSION: TDM'S CONTRIBUTION TO COSMOLOGY	65
CHAPTER 8: EXPERIMENTAL VALIDATION OF TDM	**65**
THE DOUBLE-SLIT EXPERIMENT	65
DELAYED-CHOICE EXPERIMENTS: STRENGTHENING THE CASE FOR TDM	66
THE AHARONOV–BOHM EFFECT	67
PHASE SHIFTS WITHOUT DIRECT INTERACTION	69
INFLUENCE OF ENVIRONMENTAL FACTORS	69
ROBUSTNESS ACROSS EXPERIMENTAL CONDITIONS	69
COSMIC VOIDS AND LARGE-SCALE STRUCTURE	70
COSMIC VOIDS EXPAND FASTER THAN DENSER REGIONS	71
FILAMENTARY PATTERNS REFLECT ORGANIZED ENERGY FLOWS	71
VOID-FILAMENT INTERPLAY	72
PROPOSE EXPERIMENTS AND OBSERVATIONS	**72**
VOID EXPANSION RATES	72
FILAMENT ENERGY DISTRIBUTION	73
BLACK HOLES AND EXTREME TIME DILATION	74
TIME DILATION IN GPS SYSTEMS	74
THERMAL EFFECTS ON QUANTUM SYSTEMS: VALIDATING THE TWILIGHT DIMENSION MODEL	75
REDUCED INTERFERENCE PATTERNS	76
LOSS OF QUANTUM ENTANGLEMENT	77

CHAPTER 9: IMPLICATIONS FOR COSMOLOGY AND PHILOSOPHY — 78

Implications for Cosmology and Philosophy — 78
Redefining Time, Space, and Causality — 78
Impact on Cosmology — 78
Applications Beyond Physics — 79
Understanding Consciousness — 79
Memory and State Activation — 79
Decision-Making as State Selection — 80
Treatments for Neurological Disorders — 80
Brain-Machine Interfaces — 80
Noise and Signal Clarity — 81
Information Loss and Entropy — 82
Error Correction and Redundancy — 82
Cryptography and Secure Communication — 82
Artificial Intelligence and Information Processing — 83
Enhancing Data Storage and Retrieval — 83
Securing Digital Communications — 83
Accelerating AI Development — 84
Rethinking Decision-Making in AI — 84
Enhancing Machine Learning Through Turbulence — 85
Addressing AI Bias with Resistance Modeling — 85
Building Adaptive Systems — 85
Improving AI Roustness and Resilience — 86

CHAPTER 10: FUTURE DIRECTIONS AND OPEN QUESTIONS — 87

Conclusion — 89

CHAPTER 11: PROVISIONAL PATENT APPLICATIONS EXAMPLES — 91

Provisional Patent Application: Twilight Dimension Model-Based Neural State Activation System — 91
Provisional Patent Application: Dynamic Consciousness Activation Framework — 94
Provisional Patent Application: Neural Energy Flow Modulation System for Advanced Diagnostic and Therapeutic Applications — 97
Provisional Patent Application: Twilight Dimension-Based Error Correction System — 100
Provisional Patent Application: TDM-Based Cryptographic System for Enhanced Data Security — 103
Provisional Patent Application: Adaptive Machine Learning and Communication System Based on Twilight Dimension Dynamics — 106
Provisional Patent Application: Twilight Dimension Model-Based Cryptography for Enhanced Data Security — 109
Provisional Patent Application: Enhancing Machine Learning with Turbulence Dynamics from the Twilight Dimension Model (TDM) — 112
Provisional Patent Application: Resistance-Based Bias Mitigation in AI Systems Inspired by the Twilight Dimension Model (TDM) — 115
Provisional Patent Application: Adaptive Artificial Intelligence Systems Inspired by Twilight Dimension Model (TDM) — 118
Provisional Patent Application: Turbulence-Based AI for Enhanced Search and Recommendation Systems — 121
Provisional Patent Application: AI Systems for Personalized Healthcare Using Resistance Modeling — 123

Chapter 1: Introduction to the Twilight Dimension Model

The Twilight Dimension Model (TDM) is a theoretical framework that proposes the existence of a timeless, static "twilight dimension," where all possible states of reality reside. This model posits that observable phenomena in the "reality dimension" emerge from interactions between these two dimensions. TDM aims to bridge the gaps between quantum mechanics, general relativity, and cosmology, providing a unified explanation for phenomena like wave-particle duality, dark energy, and time. This chapter introduces TDM, outlines its fundamental principles, and establishes its importance in addressing unanswered questions in physics.

The Need for a Unifying Framework
Physics today is fragmented between the quantum and macroscopic worlds.

Quantum mechanics explains the behavior of particles at the smallest scales, with phenomena like wave-particle duality, superposition, and entanglement defying classical intuition. General relativity, by contrast, excels in describing the curvature of space time, gravity, and the motion of celestial bodies.

However, these two pillars of modern physics remain fundamentally incompatible—while quantum mechanics operates probabilistically, relativity is deterministic and continuous. The inability to reconcile these frameworks not only leaves our understanding incomplete but also restricts the potential applications of a unified theory in both theoretical and practical domains.

For instance, the double-slit experiment, a cornerstone of quantum mechanics, vividly illustrates the divide. The experiment demonstrates wave-particle duality, where particles like electrons, photons or even small molecules behave as waves when unobserved, forming interference patterns, yet act like particles when observed.

Quantum mechanics describes this phenomenon mathematically using wave functions, but the deeper "why" behind this behavior remains elusive. A unifying framework like the TDM has the potential to go beyond describing the phenomenon to fundamentally explain its cause, connecting it to broader principles that link the quantum and macroscopic worlds.

The implications of such a framework extend far beyond theoretical curiosity. If we could fully understand and control the mechanisms underlying quantum phenomena like those in the double-slit experiment, it would revolutionize technologies reliant on quantum mechanics, such as quantum computing, cryptography, and advanced materials. For example, understanding how observation collapses quantum states could lead to more efficient quantum information processing, reducing decoherence in quantum computers. Additionally, linking quantum mechanics with gravity through a unified framework could unlock breakthroughs in energy technologies, space exploration, and even our understanding of the universe's origin.

Ultimately, a unifying theory like TDM addresses not only fundamental questions but also practical challenges, providing the intellectual scaffolding to advance human knowledge and capabilities. By reconciling quantum mechanics, relativity, and cosmology, such a framework promises to transform both our theoretical understanding and the technologies that shape our future.

Existing Theories and Their Explanations

Physics has provided us with a remarkable set of tools and theories to describe the universe, from the tiniest particles to the vast cosmos. These theories have reshaped our understanding of reality and led to groundbreaking technologies, yet each comes with limitations. A more unified framework is necessary to address the unanswered questions that arise when these theories are applied outside their domains.

Quantum mechanics stands at the forefront of modern physics, offering a detailed mathematical framework for describing the behavior of particles at the smallest scales. At the heart of quantum mechanics are the contributions of Werner Heisenberg and Erwin Schrödinger. Heisenberg's uncertainty principle states that we cannot simultaneously know a particle's exact position and momentum. This limitation is not due to flaws in measurement technology but is a fundamental property of nature. Schrödinger, meanwhile, introduced the **wave function**, a mathematical description of the probability of a particle's position and state. His famous thought experiment, Schrödinger's cat, highlights the strange quantum reality where a system can exist in multiple states—alive and dead—until observed. These ideas have revolutionized our understanding of the microscopic world, explaining phenomena like the behavior of electrons in atoms and the probabilistic nature of quantum states.

However, while quantum mechanics provides powerful tools to predict particle behavior, it struggles to answer the "why" behind these phenomena. For example, in the double-slit experiment, electrons act like waves when unobserved but collapse into particle-like behavior upon measurement. Heisenberg's and Schrödinger's frameworks describe this behavior mathematically but do not explain what causes the wave function to collapse.
This gap has profound implications for our understanding of reality and limits our ability to fully harness quantum mechanics for applications like quantum computing, where maintaining superposition and coherence is essential.

On the macroscopic scale, Einstein's **general relativity** offers a beautifully simple explanation of gravity. Rather than seeing gravity as a force, relativity shows it as the curvature of space time caused by mass and energy. This theory has allowed us to predict phenomena like black holes, gravitational waves, and the bending of light around massive objects. It also explains time dilation—how time moves more slowly in stronger gravitational fields or at higher velocities, a principle that underpins technologies like GPS. Yet, general relativity breaks down at quantum scales, such as near black hole singularities, where the fabric of space time itself becomes undefined.

Additionally, relativity provides no insight into the nature of **dark energy**, the mysterious force driving the accelerated expansion of the universe.

Complementing these theories is the **Standard Model of particle physics**, which catalogs the fundamental particles and forces of nature. It describes three of the four known forces—electromagnetic, weak, and strong nuclear forces—through particles like photons, gluons, and W and Z bosons. The discovery of the Higgs boson in 2012 validated the mechanism by which particles gain mass, marking a triumph for the Standard Model. However, the model excludes gravity, leaving it incomplete. It also cannot account for **dark matter**, an invisible substance that makes up most of the universe's mass, or the asymmetry between matter and antimatter in the universe's early moments.

Efforts to bridge these gaps have given rise to new theories. **String theory** proposes that particles are not point-like but are instead tiny vibrating strings, with the vibrations determining the particle's properties. It introduces additional dimensions beyond the familiar three of space and one of time, which could provide a unified explanation for all forces, including gravity. However, string theory remains untested due to the extreme energies required to probe its predictions.

Similarly, loop quantum gravity attempts to quantize space time itself, suggesting that space and time are made of discrete units. While promising, it has not yet produced predictions that can be experimentally verified.

These theories and models have significant implications for modern technology. Quantum mechanics underpins advancements like semiconductors, lasers, and quantum computers, but its unresolved mysteries limit the scalability of quantum technologies. General relativity is essential for technologies like satellite navigation and astrophysical modeling, yet its inability to incorporate quantum effects restricts its application to the very early universe and black holes.

The Standard Model explains particle interactions with incredible precision, yet its failure to include dark matter or gravity leaves critical questions unanswered about the cosmos.

A unified framework is needed to integrate these theories, addressing phenomena like wave-particle duality, the nature of dark energy, and the interplay between quantum mechanics and gravity. Such a framework would not only deepen our theoretical understanding but also revolutionize technologies that rely on quantum and relativistic principles.

The TDM aspires to achieve this by providing a cohesive explanation for the emergence of time, space, and matter from interdimensional interactions, linking the quantum and cosmic scales in a single, unified vision.

Establishing TDM as a Novel Approach

The TDM offers a transformative perspective on the nature of reality by introducing the concept of the twilight dimension—a timeless, static realm that contains all possible states of existence. Unlike traditional theories, which treat time, space, and fundamental forces as inherent aspects of the universe, TDM views these properties as emergent phenomena arising from interactions between the twilight dimension and the reality dimension. In this framework, the observable universe is shaped by the flow of energy between these two dimensions, with "state activation" as the key mechanism driving the transition from potential to actual states.

At its core, TDM proposes that all possible states of matter, energy, and existence reside in the twilight dimension as latent possibilities. These states remain non manifested until energy flows from the twilight dimension into the reality dimension, activating specific states and making them observable. This process is influenced by environmental factors such as mass, velocity, and observation, which shape the energy flow and determine the sequence in which states are activated.
Through this interaction, time, space, gravity, and other phenomena emerge as dynamic properties rather than static absolutes.

TDM provides an elegant explanation for some of the most perplexing phenomena in physics. For example, in the double-slit experiment, the interference pattern arises from the interaction of non-activated states in the twilight dimension, which create turbulence in the energy flows. Observation collapses these non-activated states into a single, particle-like manifestation in the reality dimension, resolving the wave-particle duality puzzle.

Similarly, TDM explains the accelerated expansion of the universe as a result of continuous state activation, where energy injected into the reality dimension manifests as what we perceive as dark energy.

This novel approach not only bridges the divide between quantum mechanics and relativity but also extends our understanding of cosmological phenomena. By redefining foundational properties like time and space as emergent, TDM unifies disparate aspects of physics into a cohesive framework. It opens the door to new experimental possibilities, such as probing interdimensional energy flows, and offers profound implications for how we perceive and interact with the universe. In establishing TDM as a unifying theory, this model challenges long-held assumptions and invites exploration into the very fabric of reality.

Chapter 2: Structure and Dynamics of the Twilight Dimension

The twilight dimension is described as a static realm comprising all potential states arranged in nested spheres. These states remain un-activated until energy flows from the twilight dimension into the reality dimension, where they manifest as observable phenomena. The dynamics of this interaction are governed by "interdimensional turbulence," a term that captures the chaotic flows of energy between the dimensions. This chapter explores the structure of the twilight dimension, the mechanisms of its interaction with the reality dimension, and its implications for understanding the nature of existence.

The Twilight Dimension Model Explained with a Simple Analogy

Imagine a girl standing in a room, facing a wall. She holds a flashlight in her hand, and the wall is covered with countless images layered on top of one another, all invisible until illuminated. These images represent all possible states of existence in the **twilight dimension**, a timeless and static realm where every conceivable possibility exists. The flashlight in the girl's hand symbolizes the energy flow that brings these hidden images into view, activating specific states and making them real in the **reality dimension**.

When the girl turns on the flashlight and points it at the wall, only one image lights up at a time. This act of illumination represents **state activation**, where a potential state in the twilight dimension transitions into the observable reality dimension. The image she chooses to illuminate depends on her position, the direction of the flashlight, and even the brightness of the beam. Similarly, in TDM, factors like observation, environmental conditions, and energy flows determine which states are activated and experienced in the reality dimension.
Now, imagine the girl moves the flashlight across the wall. As the beam shifts, new images are illuminated, creating the sense of a sequence—one image following another. This movement represents the **emergence of time** in TDM. Although time doesn't exist as an inherent property in the twilight dimension, the sequential activation of states in the reality dimension creates the perception of a forward-moving flow.

But the wall isn't smooth. It's textured with bumps and grooves, and the flashlight beam occasionally scatters or bends. This scattering represents **interdimensional turbulence**, the interaction of multiple un-activated states within the twilight dimension. In quantum mechanics, this turbulence is observed as phenomena like interference patterns in the double-slit experiment.

When the girl focuses her flashlight sharply, she suppresses the scattering, much like how observation in quantum experiments collapses potential states into a single, measurable outcome. The girl's flashlight beam can also vary in intensity. A dim beam may only partially illuminate an image, while a bright one makes it fully visible. This corresponds to weak and strong measurements in quantum physics, where partial observation can result in mixed or hybrid behaviors.

Finally, imagine that the girl is not alone. Other people are in the room, each with their own flashlight, illuminating their sections of the wall. The images they reveal may overlap with hers or remain entirely separate. Their combined light creates patterns and flows of energy across the wall, influencing the images that appear next. This interaction reflects how the twilight dimension connects all **reality dimensions**, allowing actions in one reality to influence others indirectly through the shared potential states.

In this analogy, the girl, her flashlight, and the wall together represent the core elements of the TDM. The twilight dimension holds all possibilities (the images), while the reality dimension emerges as specific possibilities are activated (the illuminated images). Energy flows (the flashlight beam) drive this activation, giving rise to time, space, and observable phenomena. This simple scenario captures the essence of TDM, offering an accessible way to understand its intricate workings.

Overview of the Twilight Dimension Model (TDM)

The **Twilight Dimension Model (TDM)** is a theoretical framework that proposes the existence of a timeless, static realm called the **twilight dimension**, which interacts with the observable **reality dimension** to create the universe as we experience it. TDM offers a novel explanation for fundamental phenomena such as time, space, gravity, and quantum behavior, suggesting they are emergent properties arising from the interplay between these two dimensions.

The Twilight Dimension

The twilight dimension is conceptualized as a timeless and unchanging repository of all possible states of reality. These states exist as "potentialities" rather than actualized events or objects. Imagine it as a vast database or palette of every conceivable configuration of matter, energy, and existence. These potential states are organized in a structure resembling nested spheres, each representing different layers of possibilities.

The Reality Dimension

The reality dimension is the dynamic, observable universe where these potential states are sequentially activated and brought into existence. In this dimension, we experience the flow of time, the expansion of space, and the interaction of physical forces. According to TDM, these properties are not fundamental but arise from the activation of states within the twilight dimension.

Mechanism of Interaction

The interaction between the twilight and reality dimensions occurs through a process called **state activation**. Energy flows from the twilight dimension into the reality dimension, selectively activating specific states. This activation determines what is observed and experienced in the reality dimension. Factors such as observation, mass, velocity, and environmental conditions influence the energy flow and the sequence of state activation.

Emergence of Time and Space

Time, in TDM, is an emergent property resulting from the sequential activation of states. As each state is activated, it creates a sense of progression, which we perceive as the flow of time. Space emerges as the arrangement of activated states, forming the structure within which physical objects exist and interact. Both time and space are therefore dynamic and context-dependent, shaped by the interaction between the dimensions.

Quantum Behavior and Turbulence

TDM provides a unique explanation for quantum phenomena. In experiments like the double-slit, the interference patterns arise from **interdimensional turbulence**, where multiple un-activated states in the twilight dimension interact dynamically. Observation collapses this turbulence, activating a single state in the reality dimension, resolving the wave-particle duality seen in quantum mechanics.

Cosmological Implications

On a larger scale, TDM explains the accelerated expansion of the universe as the result of continuous state activation. The energy injected into the reality dimension manifests as dark energy, driving the expansion of space. Regions like cosmic voids, which expand more rapidly, are interpreted as areas with lower resistance to state activation.

Philosophical and Experimental Potential

TDM challenges conventional assumptions by redefining time, space, and causality as emergent rather than fundamental. It also opens new avenues for experimentation. For example, measuring how environmental factors like electromagnetic fields or gravity affect interference patterns could provide evidence for the interdimensional energy flows predicted by TDM.

In summary, TDM is a bold attempt to unify quantum mechanics, relativity, and cosmology. By framing the universe as a dynamic interaction between a timeless dimension of possibilities and an observable dimension of activation, it offers a comprehensive model for understanding the nature of reality and the mechanisms driving its evolution.

The Twilight Dimension: A Realm of Possibilities

The twilight dimension is at the heart of the TDM, envisioned as a timeless, static realm that holds the potential for all possible states of reality. It is not bound by time or space as we understand them; instead, it exists as a vast repository where every conceivable configuration of matter, energy, and existence resides. These potential states are not yet actualized—they exist in a latent form, waiting to be activated and brought into the observable reality dimension.

To conceptualize this, imagine the twilight dimension as a series of **nested spheres**, each layer representing a different level of possibility. The outermost sphere might contain the simplest states, such as the basic properties of particles, while deeper spheres represent increasingly complex arrangements, such as galaxies, ecosystems, and human consciousness. Each sphere interconnects with the others, forming a cohesive structure where all states coexist, interrelated and interdependent.

This nested structure emphasizes that the twilight dimension is inherently static. Unlike the dynamic reality dimension, where time and change are fundamental, the twilight dimension holds all possibilities simultaneously, without sequence or progression. It is timeless, with no "before" or "after," as all potential states exist concurrently. The arrangement of these states in nested spheres reflects their levels of complexity and the interdimensional energy flows that connect them.

The static nature of the twilight dimension does not imply inertia or inactivity. Instead, it serves as the unchanging backdrop against which the reality dimension unfolds. When energy flows between the dimensions, specific states from the twilight dimension are activated and brought into existence in the reality dimension. This activation process is selective, guided by environmental factors such as mass, velocity, and observation, which determine which states are chosen from the infinite potential of the twilight dimension.

By describing the twilight dimension as a realm of nested possibilities, TDM provides a foundation for understanding how time, space, and other emergent properties arise. The static organization of the twilight dimension contrasts with the dynamic and ever-changing nature of the reality dimension, highlighting the interplay between the two realms as the source of observable phenomena. This structure not only explains the complexity of the universe but also offers a framework for exploring the connections between quantum mechanics, relativity, and cosmology.

Energy Flow Between the Twilight Dimension and the Reality Dimension

In the TDM, the process of creating observable phenomena is governed by **energy flows** between the twilight dimension and the reality dimension. This energy flow acts as a dynamic exchange, where latent possibilities in the twilight dimension are activated into the reality dimension, and the conditions in the reality dimension influence the flow of energy back toward the twilight dimension. This bidirectional interaction is central to the emergence of time, space, and other observable effects.

To visualize this, imagine the twilight dimension as a **wire**, carrying potential energy, and the reality dimension as a **magnet**. When a magnet moves near a wire, it generates an electric current, activating energy within the wire. Similarly, the reality dimension interacts with the static potential of the twilight dimension, "moving" through it in ways shaped by physical conditions like mass, velocity, or observation. These interactions create flows of energy that activate specific states from the twilight dimension into the reality dimension.

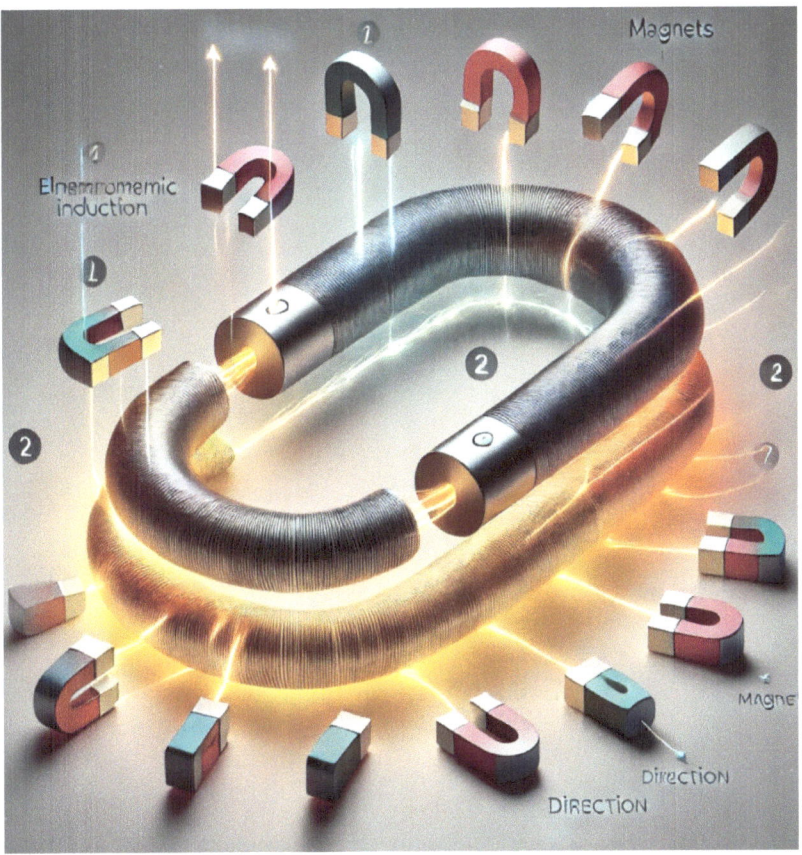

Conversely, the conditions in the reality dimension—such as the observation of a quantum system, gravitational forces, or even magnetic fields—send feedback into the twilight dimension, influencing the nature and direction of energy flows. This two-way interaction can be imagined as an intricate dance, where both dimensions continuously shape and respond to each other, creating the dynamic processes we observe as the unfolding of reality.

The flow of energy between these dimensions explains a wide range of phenomena:
Time emerges as a byproduct of sequential state activation. As energy flows from the twilight dimension into the reality dimension, states are activated one after another, creating the sense of progression we perceive as time. For example, in regions of high mass or velocity, the flow of energy slows due to increased resistance, resulting in time dilation as predicted by Einstein's relativity. Space arises from the spatial arrangement of activated states in the reality dimension. The structure and distribution of these states form the framework within which objects and phenomena exist.

Wave-particle duality in quantum mechanics is explained through interdimensional turbulence in the energy flow. In experiments like the double-slit, unobserved particles interact dynamically with turbulent flows in the twilight dimension, creating interference patterns. When observed, the turbulence collapses, and the particle manifests as a single state in the reality dimension. This flow is not static but influenced by various factors in the reality dimension. For instance:
Gravitational fields create resistance, slowing the flow of energy and altering the rate of state activation. This aligns with the observed slowing of time near massive objects.

Magnetic fields and other electromagnetic phenomena can modulate the energy flow, affecting the activation of states and potentially altering interference patterns in quantum experiments.

Observation in quantum systems represents a direct interaction, where the energy flow is aligned to select a single state from multiple potentials in the twilight dimension.

The analogy of magnets and wires also highlights the feedback loop inherent in TDM. Just as the movement of a magnet generates a current in a wire, the reality dimension's conditions generate specific flows of energy in the twilight dimension. This feedback can change the distribution or intensity of potential states, influencing future activations. For instance, the observation of a quantum system not only collapses the wave function into a single state but also alters the subsequent energy dynamics, making future states more likely to align with the observer's framework.

TDM's concept of energy flow provides a unifying explanation for both microscopic and macroscopic phenomena. On the quantum scale, it accounts for behaviors like entanglement, superposition, and wave function collapse. On the cosmic scale, it explains the accelerated expansion of the universe as the result of continuous energy injection into the reality dimension, perceived as dark energy. This ongoing activation of states expands space itself, with the feedback from the reality dimension shaping the direction and rate of expansion.

In essence, the energy flow between the twilight and reality dimensions is the engine that drives the observable universe. It links the timeless reservoir of possibilities in the twilight dimension with the dynamic, ever-changing processes of the reality dimension, creating a cohesive and interconnected model of existence. This interaction not only explains the emergence of time, space, and matter but also highlights the profound interdependence of these two realms in shaping the nature of reality.

Energy Flow and Turbulence in the Twilight Dimension Model

The TDM explains how our observable reality emerges from the interaction between two realms: the twilight dimension, a timeless, static domain containing all possible states, and the reality dimension, where those states are activated and experienced. Energy flows between these dimensions, driving the activation of states in the twilight dimension into observable phenomena in the reality dimension. This energy flow isn't smooth—it moves through nested layers, encountering resistance and turbulence along the way. To better understand this, we can use the analogy of a Tesla valve, which illustrates the directional and controlled nature of this energy flow.

Energy Flow Across the Nested Spheres

The twilight dimension can be imagined as a series of nested spheres, with each sphere representing a layer of possibilities. The simplest states, such as particles or basic properties, exist on the outer spheres, while more complex states, like molecules, ecosystems, or cosmic structures, lie deeper within.

Flow from Twilight to Reality: Energy flows outward from the twilight dimension to the reality dimension, sequentially activating states. This means that simpler states, such as particles, are often activated first, creating the foundation for more complex states like molecules or larger systems. For example, atoms must first be activated before they can combine into molecules. This process is influenced by factors like observation, gravitational fields, and velocity, which help determine which states are activated and when.

Flow from Reality to Twilight: The reality dimension sends feedback into the twilight dimension, influencing which potential states become available for activation. For example, when a quantum system is observed in the reality dimension, the act of observation affects the energy flow back into the twilight dimension. This feedback may reshape the turbulence in the twilight dimension, altering the potential states and influencing future activations.

Turbulence Over the Spheres: As energy flows across the nested spheres, it encounters turbulence, much like air moving through a storm or water flowing over rocks. This turbulence arises from the interaction of multiple potential states in the twilight dimension, competing to be activated. The effects of this turbulence include:

Delayed Activation: Some states take longer to activate because turbulence creates interference from neighboring potential states, effectively "blocking" them temporarily.

Interference Patterns: In quantum experiments like the double-slit, turbulence manifests as wave-like interference patterns. These patterns appear because un-activated states interact dynamically before observation collapses them into a single, measurable state.

Resistance to Activation: The deeper a state lies in the nested spheres (the more complex it is), the greater the resistance to activating it. Stronger or more focused energy flow is needed to bring these complex states into the reality dimension.

Tesla Valve Analogy

A Tesla valve, used in fluid dynamics, allows flow in one direction with minimal resistance while creating significant resistance in the opposite direction. This analogy helps explain the nature of energy flow in TDM:

Directional Flow: Energy flows easily from the twilight dimension into the reality dimension, activating states in a sequential order. This smooth forward flow ensures that states emerge in a structured way, creating the perception of time moving forward.

The reverse flow—from the reality dimension back to the twilight dimension—encounters more resistance. This acts as a corrective mechanism, allowing the reality dimension to influence potential states in the twilight dimension without disrupting the overall progression of activation.

Emergence of Time: Like the Tesla valve, the one-way nature of the energy flow creates a sense of time's irreversibility. States are activated in sequence, and turbulence ensures this sequence has a clear order. This directional flow underpins the emergence of time as we experience it.

Interdimensional Turbulence: Just as fluid turbulence in a Tesla valve becomes more chaotic under varying conditions, energy flowing across the nested spheres encounters turbulence. This turbulence explains phenomena like wave-particle duality, where un=activated states interact dynamically before collapsing into a single, observable state.

Key Mechanisms of Flow and Turbulence

The behavior of energy flow and turbulence is influenced by several factors:

Gravitational Resistance: Areas with strong gravitational fields (e.g., near massive objects like black holes) create resistance to energy flow. This resistance slows down state activation, aligning with the concept of time dilation, where time appears to move more slowly in regions of high gravity.

Velocity Alignment: Systems moving at high speeds relative to the energy flow alter its alignment. This creates effects similar to relativistic time contraction, where time appears to pass more quickly for fast-moving systems.

Environmental Modulation: External conditions like magnetic fields or temperature in the reality dimension can modify the energy flow. For example, magnetic fields can shift interference patterns in quantum experiments, reflecting changes in turbulence within the twilight dimension.

Observer Interaction: Observation acts like an adjustment to the Tesla valve, focusing the energy flow and suppressing turbulence. This collapses multiple potential states into a single activated state, resolving the ambiguity inherent in quantum systems and making them observable in the reality dimension.

The Role of Feedback in the Flow

The interaction between the twilight and reality dimensions isn't one-way. The energy flow from the twilight dimension activates states into reality, but the reality dimension sends feedback that reshapes turbulence and potential states in the twilight dimension. This feedback ensures a dynamic and interconnected relationship between the two dimensions, creating a self-regulating system.

The energy flow in TDM is like a carefully directed current, moving through nested spheres in the twilight dimension while encountering turbulence and resistance. The Tesla valve analogy illustrates how this flow is controlled, with a smooth forward direction and a resistant backward influence. These flows and their interactions explain the emergence of time, space, and quantum behavior, unifying phenomena across scales in a way that bridges the gaps between quantum mechanics, relativity, and cosmology.

Interdimensional Turbulence: A Bridge Between Chaos and Order

Interdimensional turbulence in the TDM refers to the chaotic flows of energy that occur as states transition between the twilight and reality dimensions. This turbulence arises from the interaction of multiple un-activated states within the twilight dimension, which compete and interfere with one another before a specific state is selected and activated. These chaotic flows are not random; they follow patterns shaped by factors such as observation, environmental conditions, and the inherent resistance of the nested spheres in the twilight dimension. The result is a dynamic and unpredictable energy landscape that influences both quantum and cosmological phenomena.
In quantum mechanics, turbulence manifests in phenomena like wave-particle duality, where interference patterns in the double-slit experiment arise from the interaction of potential states. These patterns are resolved into particle-like behavior only when an observer intervenes, collapsing the turbulence into a single outcome. On a cosmological scale, turbulence may contribute to the uneven expansion of the universe, with regions like cosmic voids experiencing faster expansion due to lower resistance in their energy flows. This concept of interdimensional turbulence provides a unifying explanation for seemingly unrelated phenomena, revealing a fundamental connection between the quantum world and the structure of the cosmos.

Chapter 3: Emergent Properties in the Twilight Dimension Model

TDM explains fundamental properties like time, space, gravity, and magnetism as emergent phenomena resulting from state activation. Time arises from the sequential activation of states, while space is defined by the arrangement of activated states. Gravity and magnetism are viewed as byproducts of directional energy flows and resistance to activation. This chapter compares TDM's emergent property framework with traditional physics, illustrating how TDM provides a deeper understanding of these phenomena through interdimensional interactions.

Emergence of Fundamental Properties in the Twilight Dimension Model

The TDM proposes that properties such as time, space, gravity, magnetism, and others are not inherent aspects of the universe but instead **emerge** as a result of **state activation**—the process by which potential states in the twilight dimension transition into observable phenomena in the reality dimension. These emergent properties are dynamic, shaped by the energy flow between the dimensions and the conditions under which states are activated.

Time: In TDM, time arises from the sequential activation of states from the twilight dimension. Each activation represents a distinct "moment," creating the perception of a linear progression. The flow of energy ensures this activation follows a defined order, giving time its directional quality. Environmental factors like mass and velocity modulate this flow, leading to phenomena such as **time dilation**, where time appears to pass more slowly near massive objects or for systems moving at high velocities. Thus, time is not a preexisting entity but a byproduct of the interplay between dimensions.

Space: Space emerges as the structural arrangement of activated states in the reality dimension. When states are activated, they create a spatial framework within which objects and phenomena can exist and interact. The dimensions and distances we perceive are shaped by the relative positions of these activated states. Unlike traditional physics, where space is a static backdrop, TDM treats space as a dynamic property, continuously redefined by the energy flow and state activation processes.

Gravity: Gravity, in TDM, is a consequence of energy flow dynamics. Mass acts as a source of resistance to the flow of energy, creating distortions in the activation process. These distortions correspond to what we observe as gravitational effects, aligning with Einstein's general relativity, where mass curves space-time. However, in TDM, this curvature is not fundamental but emergent, arising from how energy flows encounter resistance while transitioning between dimensions.

Magnetism: Magnetism is understood in TDM as a result of directional energy flows during state activation. Magnetic fields represent aligned flows of energy that influence the activation sequence of nearby states. This directional property arises naturally in systems with high degrees of order, such as those involving charged particles in motion. Magnetism, like gravity, is an emergent force shaped by the underlying interdimensional interactions.

Other Properties: Other fundamental properties, such as momentum, entropy, and light, also emerge from state activation: Momentum reflects the directional energy flow that activates states in a particular sequence, giving objects a perceived "motion." Entropy emerges from the distribution and arrangement of activated states, with increasing entropy corresponding to more diffuse or disordered activations over time. Light arises from energy transitions between states, with photons representing the direct activation of states related to electromagnetic phenomena.

Interconnected Emergence

These properties are deeply interconnected in TDM. Time and space are shaped by the sequence and arrangement of state activation, while gravity and magnetism influence how energy flows between the dimensions. For example, gravitational fields increase resistance in the flow of energy, affecting the progression of time, while magnetic fields modulate directional flows, influencing spatial arrangements and momentum. This interconnectedness highlights the unified nature of the emergent properties, offering a cohesive explanation for phenomena spanning quantum mechanics and cosmology.

By reimagining these properties as emergent rather than fundamental, TDM bridges gaps between existing theories and provides a framework for understanding how the universe's most essential aspects arise from the interaction of dimensions. This perspective not only deepens our understanding of reality but also suggests new avenues for exploring the mechanisms that shape the cosmos.

Aspect	TDM Explanation	Current Experiments	Needed
Observation Role	Actively shapes interdimensional energy flow and state activation.	Confirmed through delayed-choice and quantum eraser experiments.	Need direct measurement of energy changes during observation.
Wave-Particle Duality	Emergent from turbulence in twilight dimension flows.	Observed, but turbulence mechanism is not tested.	Experiment needed to probe environmental influences on interference patterns.
Causality	Emergent from sequential state activation.	Supported by retro-causality in delayed-choice experiments.	Requires clarification on how macroscopic causality arises from quantum activation processes.
Energy Flow	Drives state activation between dimensions.	Implied by delayed outcomes and observation effects.	No direct experimental evidence for energy flow dynamics between dimensions.
Interdimensional Turbulence	Interference patterns caused by multi-state interactions in the twilight dimension.	Interference patterns observed in standard double-slit experiments.	No experiments specifically test for turbulence-like mechanisms.
State Activation Resistance	Explains time dilation and wave function collapse.	No explicit experiments test for resistance in state activation.	Requires ultra-precise timing and energy measurements tied to environmental variables like gravity or velocity.

Comparison of the TDM to Classical Physics, Quantum Mechanics, and General Relativity

The TDM offers a unified framework that seeks to integrate and expand upon the explanations provided by classical physics, quantum mechanics, and general relativity. While each existing theory has been remarkably successful within its domain, TDM builds on these foundations by addressing gaps and providing a coherent explanation for phenomena that have remained elusive.

Below, we compare TDM's approach to the established frameworks, highlighting its integrative power:

Classical Physics vs. TDM

Classical physics, rooted in Newtonian mechanics, explains the motion of macroscopic objects through laws of force, inertia, and gravity. It assumes time and space as absolute, fixed backdrops within which objects interact. **Key Limitations of Classical Physics:** Classical physics fails at the quantum scale, where particles exhibit behaviors like superposition and wave-particle duality, and at cosmic scales, where space-time curvature influences motion. It cannot address phenomena like dark energy or time dilation.

TDM's Contributions: TDM reframes time and space as emergent properties arising from energy flows and state activation. While classical physics treats gravity as a force, TDM sees it as resistance to energy flow during state activation. By introducing the concept of interdimensional turbulence, TDM explains phenomena like interference patterns, which classical physics cannot account for. In essence, TDM replaces the static, mechanical view of classical physics with a dynamic, interdimensional perspective.

Quantum Mechanics vs. TDM

Quantum mechanics excels at describing the behavior of particles at microscopic scales, introducing principles like wave-particle duality, superposition, and entanglement. It explains probabilities through wave-functions and uses observation to collapse these probabilities into measurable outcomes.

Key Limitations of Quantum Mechanics: While quantum mechanics describes phenomena like wave-particle duality mathematically, it does not explain *why* the wave-function collapses or what fundamentally drives the probabilistic nature of particles. It also struggles to integrate gravity, as seen in the incompatibility between quantum mechanics and general relativity.

TDM's Contributions: TDM provides a physical mechanism for phenomena like wave-particle duality and wave-function collapse. It attributes interference patterns to interdimensional turbulence in the twilight dimension, where un-activated states dynamically interact. Observation aligns energy flows, collapsing turbulence into a single activated state in the reality dimension. This goes beyond mathematical description, offering a causal explanation for quantum behaviors. Additionally, TDM's inclusion of gravity as an emergent property makes it inherently compatible with quantum mechanics.

General Relativity vs. TDM

General relativity explains gravity as the curvature of space-time caused by mass and energy, providing an elegant framework for understanding large-scale phenomena such as black holes, gravitational waves, and the expansion of the universe.

Key Limitations of General Relativity: Relativity breaks down at quantum scales, where space time ceases to behave as a smooth continuum. It cannot address the nature of dark energy or unify with quantum mechanics, leaving gaps in understanding the universe's most fundamental forces.

TDM's Contributions: TDM redefines gravity not as a curvature of space-time but as resistance to energy flows during state activation. It links this resistance to time dilation, offering a unified explanation for both macroscopic and microscopic phenomena. Moreover, TDM explains dark energy as a result of continuous state activation from the twilight dimension, injecting energy into the reality dimension and driving the universe's accelerated expansion. Unlike relativity, which treats time and space as preexisting, TDM posits that these emerge dynamically from the interdimensional energy exchange.

Integration of Disparate Phenomena

One of TDM's most significant strengths lies in its ability to integrate phenomena that appear disparate under existing frameworks. For example:

Time: Classical physics treats time as absolute; relativity makes it relative to space-time; and quantum mechanics struggles to define it at all. TDM unifies these views by defining time as an emergent property of sequential state activation, reconciling its relative nature with its fundamental link to energy flows.

Wave-Particle Duality: Quantum mechanics describes it but offers no mechanism. TDM attributes it to interdimensional turbulence, where un-activated states interact dynamically before observation resolves them into a single state.

Dark Energy: General relativity attributes cosmic acceleration to an unknown force, while TDM sees it as the natural outcome of state activation, where continuous energy injection expands the fabric of space.

Gravity and Magnetism: In classical physics, these are separate forces; relativity ties gravity to space-time curvature. TDM unifies them as emergent properties of directional energy flows, influenced by resistance and turbulence during state activation.

While classical physics, quantum mechanics, and general relativity provide invaluable insights, their inability to fully describe phenomena like wave-particle duality, time, and dark energy highlights the need for a unifying framework. TDM not only bridges the gaps between these theories but also introduces new mechanisms—such as state activation and interdimensional turbulence—that provide causal explanations for longstanding mysteries. By integrating the microscopic, macroscopic, and cosmic scales, TDM offers a transformative model for understanding reality.

Chapter 4: Mechanisms of State Activation

State activation is the process by which potential states in the twilight dimension are selected and transitioned into the reality dimension. This transition depends on energy flow, environmental factors, and resistance. The Tesla valve metaphor is used to describe how resistance shapes the directionality of time, ensuring that states are activated sequentially. Observation also plays a crucial role, influencing the energy alignment required for state activation. This chapter details the mechanics of this process and its role in shaping reality.

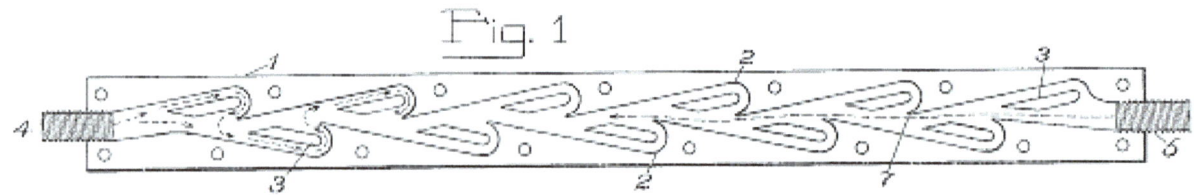

Mechanisms of State Activation in the Twilight Dimension Model

The TDM introduces the concept of **state activation** to explain how the static, timeless possibilities of the twilight dimension become the dynamic, observable phenomena of the reality dimension. In this framework, the twilight dimension serves as a vast repository of all potential states—configurations of matter, energy, and events—while the reality dimension is the stage where these potentials are brought into existence. The process of state activation bridges these two dimensions, transforming latent possibilities into measurable reality.

The twilight dimension can be imagined as a limitless library filled with books, where each book represents a possible state of the universe. These books are not being "read" or activated until energy flows from the twilight dimension into the reality dimension. This energy flow acts as a bridge, selectively transferring certain states while leaving others dormant. The act of activating these states gives rise to the properties and phenomena we observe, such as time, space, and matter.

State activation occurs in three main steps. First is **selection**, where specific states are chosen from the twilight dimension based on environmental conditions and external influences. Simpler states, such as the existence of individual particles, are often activated before more complex states like molecules or systems. This sequential selection creates the foundation for building up the observable structures of reality. Second is energy transfer, where energy flows from the twilight dimension to activate the selected states. This transfer is dynamic, requiring energy to overcome resistance, particularly for more complex states. Finally, the realization of the state occurs, where the potential state transitions fully into the reality dimension, becoming observable and measurable.

The process of state activation is not random but influenced by several factors. Observation, for example, plays a crucial role by focusing the energy flow. In quantum mechanics, this is seen as the collapse of the wave function: when a system is observed, the turbulence in the twilight dimension is resolved, and a single potential state is activated into reality. Mass and gravity also influence activation by creating resistance in the energy flow. Larger masses slow the activation process, explaining why time appears to move more slowly near massive objects, a phenomenon known as time dilation in general relativity. Velocity adds another layer of complexity, as systems moving at high speeds alter the alignment of energy flows, producing relativistic effects like time contraction.

This process is further shaped by environmental conditions such as temperature and electromagnetic fields. For instance, in quantum experiments, magnetic fields can alter the activation of states, impacting observable properties like interference patterns. The transition from potential to reality is rarely smooth, as energy flows encounter turbulence when crossing the nested spheres of the twilight dimension. Turbulence occurs when multiple potential states interact dynamically, causing delays or interference before a state can be fully activated. This turbulence is key to explaining phenomena like wave-particle duality, where particles behave like waves due to the interaction of un activated states.

The emergent properties of time, space, and matter arise directly from the mechanisms of state activation. Time emerges as the sequential activation of states, creating the perception of forward motion and progression. Space is the structural framework formed by the arrangement of activated states, providing a canvas for physical objects and interactions. Matter and energy are the activated states themselves, manifesting as observable phenomena governed by the dynamics of the energy flow.

State activation offers profound insights into both quantum mechanics and cosmology. It provides a causal explanation for wave-particle duality and wave function collapse while also addressing larger mysteries like dark energy and the expansion of the universe. By linking the timeless reservoir of possibilities in the twilight dimension with the dynamic experiences of the reality dimension, TDM provides a unified and intuitive framework for understanding how the universe transitions from potential to reality. This process underpins all observable phenomena, connecting the smallest particles to the largest cosmic structures in a single, cohesive model.

Factors Influencing State Activation in the Twilight Dimension Model
In the TDM, state activation describes the process by which potential states in the timeless twilight dimension become observable phenomena in the reality dimension. This process is governed by several factors that influence which states are activated, how quickly they transition,

and the nature of their manifestation. These factors—energy flow, resistance, and environmental conditions—work together to shape the dynamic interplay between the two dimensions.

Energy Flow

Energy flow is the primary mechanism driving state activation. It acts as a bridge between the twilight and reality dimensions, transferring potential states into observable ones. This flow can be imagined as a current moving through the nested spheres of the twilight dimension, selectively activating states along the way.

The energy flow is directed by external influences, such as observation, velocity, or gravitational forces. It determines which potential states are chosen from the twilight dimension and how they appear in the reality dimension. For instance, in quantum experiments like the double-slit, energy flows dynamically across multiple potential states, creating interference patterns. Observation focuses the flow, resolving turbulence and selecting a single state for activation.

The intensity and directionality of the energy flow are crucial. High-intensity flows can overcome resistance more easily, activating deeper, more complex states within the nested spheres. Conversely, low-intensity flows may only activate simpler states closer to the outer layers. This dynamic explains why simpler phenomena, such as particle behavior, are more readily observed than complex systems, which require stronger or more precise energy flows to manifest.

Resistance: Resistance occurs naturally during state activation and is a defining factor in the process. It arises from the structure of the nested spheres and the inherent complexity of the states being activated. States that lie deeper within the twilight dimension (i.e., more complex states) face greater resistance during activation, requiring stronger or more directed energy flows to transition into the reality dimension.

Resistance also increases due to external conditions, such as gravitational fields or mass. In regions with high gravitational mass, the energy flow encounters more resistance, slowing the activation process. This effect aligns with the concept of time dilation in general relativity, where time appears to move more slowly near massive objects. From the perspective of TDM, this slowing occurs because state activation is delayed by the resistance created by the gravitational field.

Similarly, velocity influences resistance. Systems moving at high speeds relative to the energy flow alter its alignment, creating resistance in the activation process. This corresponds to relativistic time contraction, where time appears to pass more quickly for fast-moving systems. Resistance, therefore, not only shapes the timing of activation but also impacts the sequence in which states transition from potentiality to reality.

Environmental Conditions

The surrounding environment in the reality dimension plays a significant role in modulating energy flows and shaping the state activation process. Environmental factors introduce variability into the turbulence and resistance of the energy flow, influencing how states manifest.

Magnetic Fields: Magnetic fields can alter the alignment of energy flows, affecting the activation process. For example, in quantum systems, electromagnetic fields may shift the interference patterns seen in experiments, providing indirect evidence of their influence on turbulence and state activation. Strong magnetic fields could, theoretically, direct energy flows to favor specific states, offering a tool to manipulate quantum systems.

Temperature: Higher temperatures increase thermal agitation, introducing additional turbulence into the energy flow. This can disrupt the coherence of state activation, making it harder for states to fully transition. This phenomenon explains why quantum coherence—where particles maintain superposition—is more stable at lower temperatures. Conversely, lower temperatures reduce turbulence, allowing for more precise activation.

Observation: Observation is a unique environmental factor in TDM, acting as a direct modulator of energy flow. When a system is observed, the energy flow is concentrated, collapsing turbulence and selecting a single state for activation. This explains wave function collapse in quantum mechanics, where observation resolves uncertainty into a defined outcome.

Gravitational Fields: As noted, gravitational fields create significant resistance, particularly in areas of high mass. This resistance slows the energy flow, delaying activation and altering the rate at which states transition. This effect not only explains time dilation but also suggests that gravitational influences could shape the distribution of activated states, influencing large-scale cosmic phenomena like the expansion of the universe.

Interplay of Factors

These factors—energy flow, resistance, and environmental conditions—do not operate in isolation. Instead, they interact dynamically to shape the activation process. For instance:
In a region of high gravity and low temperature, resistance might be high, but reduced turbulence could allow for precise activation of complex states.

In a high-velocity system, resistance from motion could offset the effect of magnetic fields, altering the sequence of state activation and creating relativistic effects.

The combination of these influences creates a rich and intricate energy landscape that governs the emergence of phenomena in the reality dimension. This interplay ensures that state activation is not purely deterministic but adaptable, capable of responding to a wide range of environmental and systemic conditions.

State activation in TDM is a dynamic process shaped by energy flow, resistance, and environmental factors like gravity, temperature, and observation. These influences determine which states are activated, how quickly they transition, and the way they manifest in the reality dimension. By accounting for these factors, TDM provides a robust framework for understanding the emergence of time, space, matter, and energy, offering a unifying explanation for quantum and cosmological phenomena alike. This dynamic interaction between dimensions underscores the complexity and adaptability of the universe's fundamental mechanisms.

The Tesla Valve as a Conceptual Metaphor for Resistance in State Activation

In the TDM, **resistance** plays a crucial role in shaping how potential states from the twilight dimension are activated into observable phenomena in the reality dimension. A helpful way to understand this resistance is through the metaphor of a **Tesla valve**, a fluid dynamics mechanism designed to allow fluid to flow in one direction with minimal resistance while creating significant resistance in the opposite direction. This concept mirrors the one-way flow of energy in TDM, which drives the emergence of time and observable states while maintaining order and sequence in state activation.

Additionally, the structure of the **nested spheres** in the twilight dimension can be linked to the Tesla valve concept, as turbulence and disturbances across the spheres contribute to the resistance, further influencing the activation process.

The Tesla Valve and Unidirectional Time Flow

A Tesla valve ensures that flow in one direction encounters minimal resistance, while flow in the opposite direction is impeded by design. In TDM, the energy flow from the twilight dimension to the reality dimension follows this principle:

Forward Flow (Twilight to Reality): Energy flows smoothly when activating potential states, creating a sense of progression or "forward" time in the reality dimension. This smooth flow allows for sequential state activation, where simpler states emerge before more complex ones.

Reverse Flow (Reality to Twilight): When feedback flows back from the reality dimension to the twilight dimension, it encounters resistance, ensuring that the influence of the reality dimension does not disrupt the overall sequence of activation. This resistance maintains the unidirectional nature of time while allowing subtle adjustments to potential states.

This mechanism explains why time appears to flow irreversibly in the reality dimension. The resistance encountered by reverse energy flow prevents the reactivation of previously activated states, preserving the integrity of time's forward progression.

Nested Spheres and Turbulence as Resistance

The twilight dimension is composed of **nested spheres**, each layer representing a different level of complexity in potential states. These spheres are not smooth but textured, with their surfaces influenced by interdimensional turbulence. Turbulence arises when multiple potential states interact dynamically, creating disturbances that impede energy flow.

Surface Disturbances: Imagine the nested spheres as having textured surfaces, like ridges or waves. As energy flows across these spheres, these disturbances create resistance, similar to the effect of a Tesla valve. The more complex the state (deeper within the nested spheres), the greater the resistance due to the increased interaction between neighboring potential states.

Turbulence Build-Up: As energy flows across a sphere, the interaction between potential states generates chaotic movement. This turbulence delays or redirects the flow, introducing resistance that shapes the sequence of state activation. Turbulence on the outer spheres may manifest as quantum-level interference patterns, while deeper turbulence could influence cosmic-scale phenomena like dark energy or galaxy formation.

Integrating the Tesla Valve with Nested Spheres

The Tesla valve metaphor can be extended to the nested spheres of the twilight dimension. Each sphere acts as a valve layer, regulating the flow of energy toward the reality dimension:
Selective Pathways: Just as a Tesla valve channels fluid through specific pathways, the nested spheres guide energy flow along paths of least resistance. This selective activation ensures that simpler states are activated before more complex ones, creating a structured progression.
Cumulative Resistance: The deeper the energy flows into the nested spheres, the more cumulative resistance it encounters. Each layer adds its own "valve effect," with the textured surfaces and turbulence within the spheres compounding the difficulty of activating deeply nested states.

Directional Flow: The structure of the nested spheres ensures that energy flows outward toward the reality dimension with minimal resistance while making reverse flows (feedback) more difficult. This design preserves the forward progression of time and maintains the integrity of emergent properties like space and matter.

Implications for State Activation

The Tesla valve metaphor, combined with the concept of nested spheres, explains several key aspects of TDM:

Wave-Particle Duality: Turbulence across the outer spheres creates interference patterns, much like chaotic flows in a Tesla valve. Observation resolves these patterns, focusing the energy flow to activate a single state.

Time Dilation: Resistance from gravitational fields or high velocity increases the "valve effect," slowing the energy flow and delaying state activation. This corresponds to the slowing of time in strong gravitational fields or for fast-moving systems.

Complexity Emergence: States deeper within the nested spheres face greater resistance, requiring more focused or intense energy flows to activate. This explains why complex phenomena, like molecular interactions or cosmic structures, emerge later in the sequence of activation.

The Tesla valve serves as a powerful metaphor for understanding resistance in the state activation process of TDM. Its ability to direct energy flow while impeding reverse flows mirrors the one-way progression of time and the structured emergence of phenomena. When integrated with the concept of nested spheres, the Tesla valve analogy highlights how turbulence and surface disturbances create resistance, shaping the sequence and nature of state activation. Together, these ideas provide a cohesive framework for explaining how the twilight dimension transitions potential states into the dynamic reality we observe, preserving order while allowing for complexity to emerge.

The Relationship Between Activation and Observer Influence in the Twilight Dimension Model

In the TDM, **state activation**—the process by which potential states in the twilight dimension become observable in the reality dimension—is fundamentally shaped by the act of observation. Observation does more than simply "measure" reality; it plays an active role in determining which states are activated, influencing the energy flow between the dimensions and resolving the turbulence of un-activated possibilities. This relationship between activation and observation not only bridges quantum phenomena with larger-scale realities but also underscores the role of the observer as a co-creator of the observable universe.

Observer Influence as Energy Flow Modulation

In TDM, the twilight dimension contains all possible states, existing as latent potential within nested spheres. The reality dimension emerges as specific states from this reservoir are activated. However, the activation process is not automatic or random—it is highly influenced by observation.

Observation in TDM can be thought of as a **modulator of energy flow**. When an observer interacts with a system, they focus the energy flow from the twilight dimension to the reality dimension, collapsing the interdimensional turbulence of potential states. This focused flow aligns the energy in a way that selects and activates a single state, bringing it into observable reality. Without observation, multiple potential states coexist, interacting chaotically and creating turbulence that manifests as wave-like interference patterns in quantum systems.

For example, in the **double-slit experiment**, unobserved particles interact with turbulence in the twilight dimension, creating interference patterns that indicate the presence of multiple un-activated states. When the system is observed, this turbulence is resolved, and the energy flow aligns to activate a single state—causing the particle to behave as if it passed through only one slit.

Shaping Reality Through Observation

Observation not only resolves uncertainty but also actively **shapes the outcome of reality**. In TDM, the observer plays a critical role in determining how energy flows and which states are activated. This influence can be understood in several ways:

Collapse of Potential States: In quantum mechanics, the act of observation collapses a wave function, selecting a definite outcome from a range of probabilities. TDM extends this idea by describing how observation collapses the turbulence in the twilight dimension, aligning energy flow to activate a single state. This transforms the probabilistic potential of the twilight dimension into the deterministic reality of the observed dimension.

Feedback to the Twilight Dimension: Observation not only activates a state but also sends feedback into the twilight dimension, influencing the alignment of potential states for future activation. This feedback loop ensures that the act of observing shapes not just the present but also the possibilities for subsequent activations.

Observer's Role in Time and Space: By focusing the energy flow, observation organizes the sequence of state activation, giving rise to the emergent properties of time and space. The order in which states are activated is influenced by the observer's frame of reference, connecting observation to the perception of temporal and spatial progression.

Observer Influence in Macroscopic Phenomena

While the role of observation is well-documented in quantum mechanics, TDM extends its influence to macroscopic and cosmological phenomena. For instance:

Cosmic Evolution: Observation by conscious beings may subtly direct the flow of energy in ways that influence large-scale structures like galaxies, echoing the anthropic principle, which suggests that the universe's properties are fine-tuned for observers.

Time's Flow: Observation anchors the sequence of state activation, reinforcing the perception of forward-moving time. Without observation, turbulence might allow for random or even retroactive activation sequences, challenging the idea of causality.

Experimental Implications of Observer Influence

The observer's role in shaping reality opens exciting possibilities for experimental validation of TDM. For instance:

Delayed-Choice Experiments: These experiments demonstrate that the act of observation seems to retroactively affect which path a particle took, even after it has already passed through the experimental apparatus. TDM explains this by suggesting that observation sends feedback to the twilight dimension, realigning potential states even after they have influenced the reality dimension.

Weak Measurements: In weak measurement setups, partial observation allows some interdimensional turbulence to persist, creating hybrid behaviors that reflect both particle- and wave-like properties. This supports TDM's prediction that observation modulates energy flow without fully collapsing turbulence in these cases.

Interdependence of Observation and Reality

TDM reveals a profound interdependence between observation and reality:
The **reality dimension** depends on the twilight dimension for its reservoir of potential states.
The **twilight dimension** depends on observers in the reality dimension to guide the energy flow, resolving turbulence and selecting states for activation.

In this way, observation is not a passive act but a dynamic interaction, where the observer actively shapes the universe through their influence on state activation. This interaction redefines the observer's role, placing them as an integral participant in the unfolding of reality.
In TDM, observation is a driving force that shapes reality by influencing energy flows and resolving interdimensional turbulence. It determines which states are activated, aligns the sequence of activation, and feeds back into the twilight dimension to influence future

possibilities. This relationship highlights the active role of observers in the universe, bridging the microscopic and macroscopic scales and offering a cohesive explanation for phenomena ranging from quantum wave-function collapse to the progression of time and the formation of cosmic structures. By integrating observation into the activation process, TDM provides a comprehensive framework for understanding how reality emerges from potentiality.

Chapter 5: Time as an Emergent Property

Time, in TDM, is not a fundamental entity but an emergent property of state activation. The flow of energy from the twilight dimension activates states in a sequence, creating the perception of time. Gravitational and relativistic effects, such as time dilation, are explained as variations in resistance to state activation. Delayed-choice experiments and retro-causality are revisited through TDM, demonstrating how time is flexible and context-dependent. This chapter redefines time as a product of interdimensional energy dynamics.

Time in the Twilight Dimension Model (TDM): A Byproduct of State Activation

In the TDM, **time** is not a fundamental property of the universe but an **emergent phenomenon** that arises from the sequential activation of potential states in the twilight dimension. This activation occurs through the flow of energy between the twilight and reality dimensions. As energy flows and states are activated one after another, the perception of time emerges as a byproduct of this progression.

Sequential State Activation: The Building Blocks of Time

The twilight dimension contains all possible states of existence, arranged in nested spheres of increasing complexity. These states remain latent until activated by energy flowing from the twilight dimension into the reality dimension. The process of activation is not random but follows a structured sequence influenced by environmental factors like mass, velocity, and observation.

Each activation represents a distinct "moment," and the sequence of these moments creates the perception of time in the reality dimension. For example:
The activation of simpler states, such as fundamental particles, forms the foundation for more complex arrangements, such as molecules or systems.
As these states are activated in a defined order, the flow of energy ensures a sense of continuity and progression, which we interpret as the forward movement of time.
Without sequential activation, there would be no progression of events, and time as we experience it would not exist. This sequence is not inherent to the twilight dimension, where all states exist simultaneously, but emerges only in the reality dimension as a result of energy flow.

Interdimensional Energy Flow and the Direction of Time

The energy flow between the dimensions acts as the engine driving state activation and the emergence of time. This flow is analogous to a current, moving from the twilight dimension into the reality dimension. However, this movement is not perfectly smooth; it encounters turbulence and resistance, which influence the rate and order of activation.
The **Tesla valve analogy** helps explain the one-way directionality of this flow:
Energy flows more easily from the twilight dimension to the reality dimension, activating states in a forward sequence.

The reverse flow, where feedback from the reality dimension influences potential states in the twilight dimension, encounters resistance. This resistance prevents states from being reactivated out of order, ensuring the unidirectional nature of time.
The **irreversibility of time** is thus a natural consequence of this energy flow. The forward progression of state activation ensures that each moment is distinct and cannot be undone. Even though feedback from the reality dimension can influence future activations, it cannot reverse the sequence of previously activated states.

Environmental Factors Influencing the Flow of Time

The rate at which time progresses depends on the conditions influencing state activation. These conditions create variations in the flow of energy, altering the perception of time in different contexts:

Mass and Gravity: In regions of high mass, such as near black holes, gravitational fields create resistance to energy flow. This slows the activation of states, causing time to pass more slowly—a phenomenon known as **gravitational time dilation**. In TDM, this resistance is a direct result of the interaction between energy flows and the nested spheres of the twilight dimension.

Velocity: Systems moving at high speeds relative to the energy flow experience a shift in the alignment of activation, leading to relativistic time dilation. In such systems, the sequence of state activation is stretched, making time appear to pass more slowly for fast-moving objects.

Observation: Observation focuses and modulates the energy flow, influencing the activation process. By resolving turbulence, observation ensures the sequential activation of states, reinforcing the perception of time's forward progression.

The Emergence of Time vs. Timelessness in the Twilight Dimension
While time is a defining feature of the reality dimension, the twilight dimension itself is timeless. In the twilight dimension, all potential states exist simultaneously, unbound by progression or sequence. This timelessness provides the foundation for the emergent property of time, which arises only when energy flows select and activate states in a specific order.

This distinction highlights a key feature of TDM:
The twilight dimension is a static reservoir of possibilities, where time has no meaning.
The reality dimension is dynamic, with time emerging as a consequence of the interactions between energy flow, resistance, and state activation.

Implications for Time Perception and Reality
The emergent nature of time in TDM has profound implications for understanding reality:
Relativity of Time: Variations in state activation rates explain why time appears to flow differently in different conditions, aligning with Einstein's relativity while providing a deeper mechanism for time dilation.

Directionality: The one-way flow of energy and the resistance to reverse activation explain why time appears irreversible, aligning with the arrow of time observed in thermodynamics.

Interdimensional Feedback: While time flows forward, feedback from the reality dimension influences future state activation, showing how past and present conditions shape potential futures without violating the sequence of time.

In TDM, time is not an inherent property of the universe but an emergent feature of sequential state activation and interdimensional energy flow. It arises as energy transfers from the twilight dimension into the reality dimension, activating states in a structured progression. Influenced by environmental factors like gravity, velocity, and observation, time's flow varies in different contexts but always retains its forward direction due to the resistance inherent in reverse activation. By redefining time as a dynamic, emergent phenomenon, TDM bridges quantum mechanics, relativity, and cosmology, offering a unified framework for understanding one of the universe's most fundamental experiences

Resistance in Activation and Time Dilation in the Twilight Dimension Model
In the TDM, time dilation—the slowing of time in certain conditions—is explained as a direct consequence of resistance encountered during the activation of potential states from the twilight

dimension into the reality dimension. This resistance, influenced by factors such as gravitational fields and velocity, aligns with and extends the principles of Einstein's relativity, providing a deeper mechanistic understanding of why time flows differently in different contexts.

State Activation and Resistance

In TDM, time emerges as a result of **sequential state activation**. Each potential state in the twilight dimension must be activated by energy flow to become part of the observable reality dimension. However, this activation process is not uniform. Resistance—opposing forces that slow the flow of energy—can delay the activation of states. When resistance is high, the progression of state activation slows, causing time to appear to pass more slowly in the affected region.

Gravitational Effects and Time Dilation

Massive objects, such as planets, stars, or black holes, create **gravitational fields** that significantly increase resistance to energy flow. In these high-gravity regions:
The energy flowing from the twilight dimension encounters greater resistance due to the curvature and compression of space around massive objects.

This resistance slows the rate at which states are activated, effectively reducing the progression of time in the reality dimension near these objects.

This explanation aligns with general relativity, where time dilation occurs because space-time is distorted by mass. In TDM, the distortion is reframed as resistance within the nested spheres of the twilight dimension, which impedes the flow of energy. The greater the mass, the higher the resistance, and the slower the activation of states, leading to a corresponding slowdown in the perception of time.

For example, near a black hole, the immense gravitational field creates such high resistance that time dilation becomes extreme—an hour near the black hole might equate to years for an observer far from its gravitational influence. TDM interprets this as the near-halting of state activation due to the overwhelming resistance posed by the gravitational field.

Velocity and Time Dilation

Time dilation also occurs when objects move at high velocities relative to their surroundings. In TDM, this is explained by the **alignment of energy flows** and the increased resistance caused by motion:

At high velocities, the relative motion between the object and the energy flow alters the path of state activation, increasing the resistance encountered. This resistance slows the rate of state activation for the moving object, causing time to appear to pass more slowly for it compared to an observer at rest. This aligns with Einstein's theory of special relativity, where time dilation occurs as an object approaches the speed of light. TDM adds a mechanistic layer to this phenomenon: as velocity increases, the alignment of energy flow shifts, creating turbulence and resistance that delay activation. This not only explains time dilation but also highlights the interconnectedness of energy, motion, and state activation. For instance, an astronaut traveling at near-light speed experiences time more slowly than someone on Earth. In TDM, this is due to the increased resistance in energy flow caused by the astronaut's velocity, which slows the activation of sequential states, delaying their perception of time.

Comparing Gravitational and Velocity-Based Resistance

While both gravitational fields and velocity create resistance, their effects differ in origin and application:

Gravitational Resistance: Arises from the compression or distortion of space around massive objects, increasing resistance in all directions. This causes time to slow uniformly near the mass.

Velocity-Based Resistance: Results from the relative motion of an object, creating directional resistance due to misalignment with the natural energy flow. This causes time dilation only for the moving object.

In both cases, resistance delays the rate of state activation, demonstrating a consistent mechanism for time dilation across different physical contexts.

Implications of Resistance-Based Time Dilation

By framing time dilation as a result of resistance in state activation, TDM integrates relativity's predictions into a broader framework:

Unified Explanation: Both gravitational and velocity-based time dilation are understood as manifestations of the same underlying mechanism—resistance to energy flow during activation.
Emergence of Time: Time is not an independent entity but an emergent property shaped by the dynamic conditions influencing energy flow. When resistance increases, time appears to slow, underscoring its dependence on interdimensional interactions.

Scalability: This model bridges the quantum and cosmic scales. For example, gravitational resistance near a black hole mirrors the turbulence seen in quantum systems, suggesting a unified principle for time dilation across all scales of reality.
In TDM, time dilation is a natural consequence of resistance encountered during state activation. Gravitational fields and high velocities increase this resistance, slowing the energy flow and delaying the sequential activation of states. This resistance-based mechanism aligns with the predictions of general and special relativity while providing a deeper, unified understanding of how time emerges and behaves in different conditions. By framing time dilation as an outcome of interdimensional energy dynamics, TDM bridges gaps between classical physics, relativity, and quantum mechanics, offering a comprehensive explanation for one of the universe's most fascinating phenomena.

Gravitational Effects and Time Dilation

Massive objects, such as planets, stars, or black holes, create **gravitational fields** that significantly increase resistance to energy flow. In these high-gravity regions:
The energy flowing from the twilight dimension encounters greater resistance due to the curvature and compression of space around massive objects. This resistance slows the rate at

which states are activated, effectively reducing the progression of time in the reality dimension near these objects.

This explanation aligns with general relativity, where time dilation occurs because space time is distorted by mass. In TDM, the distortion is reframed as resistance within the nested spheres of the twilight dimension, which impedes the flow of energy. The greater the mass, the higher the resistance, and the slower the activation of states, leading to a corresponding slowdown in the perception of time.

For example, near a black hole, the immense gravitational field creates such high resistance that time dilation becomes extreme—an hour near the black hole might equate to years for an observer far from its gravitational influence. TDM interprets this as the near-halting of state activation due to the overwhelming resistance posed by the gravitational field.

Velocity and Time Dilation

Time dilation also occurs when objects move at high velocities relative to their surroundings. In TDM, this is explained by the **alignment of energy flows** and the increased resistance caused by motion:

At high velocities, the relative motion between the object and the energy flow alters the path of state activation, increasing the resistance encountered.
This resistance slows the rate of state activation for the moving object, causing time to appear to pass more slowly for it compared to an observer at rest.

This aligns with Einstein's theory of special relativity, where time dilation occurs as an object approaches the speed of light. TDM adds a mechanistic layer to this phenomenon: as velocity increases, the alignment of energy flow shifts, creating turbulence and resistance that delay activation. This not only explains time dilation but also highlights the interconnectedness of energy, motion, and state activation.
For instance, an astronaut traveling at near-light speed experiences time more slowly than someone on Earth. In TDM, this is due to the increased resistance in energy flow caused by the astronaut's velocity, which slows the activation of sequential states, delaying their perception of time.

Comparing Gravitational and Velocity-Based Resistance

While both gravitational fields and velocity create resistance, their effects differ in origin and application:

Gravitational Resistance: Arises from the compression or distortion of space around massive objects, increasing resistance in all directions. This causes time to slow uniformly near the mass.
Velocity-Based Resistance: Results from the relative motion of an object, creating directional resistance due to misalignment with the natural energy flow. This causes time dilation only for the moving object.

In both cases, resistance delays the rate of state activation, demonstrating a consistent mechanism for time dilation across different physical contexts.

Implications of Resistance-Based Time Dilation

By framing time dilation as a result of resistance in state activation, TDM integrates relativity's predictions into a broader framework:

Unified Explanation: Both gravitational and velocity-based time dilation are understood as manifestations of the same underlying mechanism—resistance to energy flow during activation. Emergence of Time: Time is not an independent entity but an emergent property shaped by the dynamic conditions influencing energy flow. When resistance increases, time appears to slow, underscoring its dependence on interdimensional interactions.

Scalability: This model bridges the quantum and cosmic scales. For example, gravitational resistance near a black hole mirrors the turbulence seen in quantum systems, suggesting a unified principle for time dilation across all scales of reality.

In TDM, time dilation is a natural consequence of resistance encountered during state activation. Gravitational fields and high velocities increase this resistance, slowing the energy flow and delaying the sequential activation of states. This resistance-based mechanism aligns with the predictions of general and special relativity while providing a deeper, unified understanding of how time emerges and behaves in different conditions. By framing time dilation as an outcome of interdimensional energy dynamics, TDM bridges gaps between classical physics, relativity, and quantum mechanics, offering a comprehensive explanation for one of the universe's most fascinating phenomena.

Time and Causality in the Twilight Dimension Model

In the TDM, time is not a fundamental construct but an emergent property of the sequential activation of potential states from the twilight dimension to the reality dimension. This sequential nature of state activation forms the foundation for causality, the principle that events are linked by cause-and-effect relationships. By defining time as the progression of activated states, TDM provides a framework for understanding how causality emerges naturally and why it is deeply intertwined with the perception of time.

State Activation and the Sequence of Events

In the twilight dimension, all possible states exist simultaneously, without a predefined order. It is only when energy flows from the twilight dimension to the reality dimension, activating states in a specific sequence, that time and causality arise. Each activated state represents a distinct "moment," and the order of these moments determines the flow of events in the reality dimension.

This sequence is inherently directional:

Forward Activation: States are activated one after another, creating the perception of time moving forward. For example, simpler states, such as fundamental particles, are activated first, forming the basis for more complex states like molecules, systems, or events.

Causal Links: The sequence of activation establishes the relationships between events. If state A is activated before state B, A is perceived as the cause of B. This relationship is a direct consequence of the order in which states are realized in the reality dimension.
Without sequential activation, there would be no distinction between cause and effect, as all states would exist simultaneously and indistinguishably.

The Role of Energy Flow in Causality

The flow of energy between the dimensions is the engine that drives this sequence, ensuring that states are activated in a structured progression. This flow:

Establishes Order: The energy flow follows a path of least resistance, activating states in an orderly fashion. This structured activation defines the "arrow of time," ensuring that earlier states influence later ones, but not the reverse.

Preserves Causality: The resistance inherent in reverse energy flow (as explained by the Tesla valve analogy) prevents states from being reactivated out of order. This resistance ensures that once a state is activated, it cannot retroactively influence prior states, preserving the integrity of cause-and-effect relationships.

For example, in a system where a ball is dropped and then hits the ground, the state representing the ball falling (cause) must be activated before the state of it hitting the ground (effect). The energy flow ensures this sequence, and the resistance prevents the effect (the ball hitting the ground) from influencing the cause (the ball being dropped).

Causality Across Scales: From Quantum to Cosmic

TDM's linking of time and causality provides insights across both quantum and cosmic scales:
In Quantum Mechanics:
In quantum systems, causality is tied to the collapse of potential states into a single observed state. Observation focuses the energy flow, resolving interdimensional turbulence and ensuring that the observed outcome aligns with the sequence of state activation. For instance, in the delayed-choice experiment, TDM explains how observation retroactively influences potential states in the twilight dimension while preserving the overall sequence of activation.

In Cosmology: On a cosmic scale, causality governs the evolution of the universe. The activation of states drives the formation of galaxies, stars, and planets in a progressive sequence. For example, the state representing the initial conditions of the universe (e.g., the Big Bang) must be activated before states representing the formation of galaxies or life. This progression is maintained by the energy flow and the resistance to reverse activation.

Causality and the Emergence of Time

Time's emergence in TDM is inseparable from causality. As states are activated in sequence, time provides the framework within which causes lead to effects:
Past, Present, and Future: Activated states form the "past," while states currently being activated represent the "present." Potential states awaiting activation are the "future." This division is what allows us to perceive a clear progression of cause and effect.

Directionality of Time: The forward direction of state activation gives time its arrow, ensuring that effects cannot precede their causes. This directionality is preserved by resistance to reverse flow, which prevents the reactivation of previously activated states.

TDM's Insights into Causality

TDM offers unique insights into causality by connecting it to the mechanics of state activation and energy flow:

Unified Framework: TDM unifies quantum mechanics and relativity by explaining causality as a function of sequential state activation. It resolves the apparent contradictions between quantum retrocausality (e.g., delayed-choice experiments) and macroscopic causality by showing that feedback influences future activations without disrupting the overall sequence.

Dynamic Interplay: Feedback from the reality dimension influences the twilight dimension, aligning potential states for future activation. This creates a dynamic system where causality emerges not just from static rules but from the active interplay of dimensions.

Implications for Time Perception: The perception of time and causality is relative to the observer's frame of reference. In regions of high resistance, such as near massive objects or for high-velocity systems, time dilation slows the activation of states, altering the observer's experience of cause and effect. This aligns with Einstein's relativity while extending it to interdimensional interactions.

In TDM, time and causality are deeply interconnected, both arising from the sequential activation of states driven by interdimensional energy flow. Time provides the framework for understanding cause and effect, while causality defines the order of state activation, ensuring a coherent progression of events. Resistance to reverse flow preserves the integrity of this sequence, preventing retroactive influences and maintaining the directionality of time. By linking time's emergence to causality, TDM provides a unified explanation for phenomena across scales, bridging the gaps between quantum mechanics, relativity, and cosmology. This framework highlights the profound role of interdimensional dynamics in shaping the universe's most fundamental experiences.

Experimental Evidence Supporting TDM's Concept of Time

The Twilight Dimension Model (TDM) redefines time as an emergent property arising from the sequential activation of states in the twilight dimension. This framework is supported by various experimental findings and observations that, while interpreted differently under traditional physics, align naturally with TDM's mechanisms of state activation, energy flow, and resistance. Below are five key experiments and phenomena that provide compelling support for TDM's view of time.

Delayed-Choice and Quantum Eraser Experiments

Key Findings: In delayed-choice experiments, a particle's behavior (wave or particle-like) appears to be influenced by measurements made after it has passed through an experimental apparatus.

Quantum eraser setups further demonstrate that information "erased" about a particle's path retroactively alters the outcome of the interference pattern.

TDM Interpretation: In TDM, these results are explained as feedback from the reality dimension influencing the alignment of potential states in the twilight dimension. Observation acts as a modulator of energy flow, collapsing turbulence in the twilight dimension and determining which state is activated into reality. The timeless nature of the twilight dimension allows for this feedback to realign states even after earlier activations, preserving the causality of state progression in the reality dimension. The experiments highlight how time and causality in TDM are emergent and adaptable, rather than fixed absolutes.

Time Dilation in General Relativity

Key Findings: Experiments such as Hafele-Keating (flying atomic clocks around the Earth) and observations of time dilation near black holes (e.g., in the movie "Interstellar," based on real calculations) confirm that time slows in strong gravitational fields or for fast-moving systems. Atomic clocks run slower at higher velocities or in stronger gravitational fields, demonstrating that time is not absolute.

TDM Interpretation: TDM explains time dilation as a resistance in state activation caused by environmental factors: Gravitational Resistance: Mass creates resistance in the flow of energy between the dimensions, delaying the sequential activation of states and slowing time.
Velocity-Induced Resistance: High velocities alter the alignment of energy flows, increasing turbulence and resistance, which delays state activation. These effects directly correlate with the experimental findings, providing a mechanistic understanding of time dilation as a product of interdimensional dynamics.

Bending of Light by Gravity (Gravitational Lensing)

Key Findings: Observations such as the bending of starlight around massive objects (first confirmed during the 1919 solar eclipse) show that light follows curved paths in the presence of strong gravitational fields. This phenomenon supports Einstein's theory of general relativity, which predicts space-time curvature caused by mass.

TDM Interpretation: In TDM, gravitational fields are interpreted as regions of high resistance in the flow of energy between dimensions. Light, which represents activated states, interacts with this resistance, altering its path. The bending of light demonstrates how activated states are influenced by the energy flow dynamics dictated by mass, reinforcing TDM's view of time as emergent from the same mechanisms. The relationship between resistance, state activation, and time aligns with the bending of light as a direct consequence of mass-induced disruptions in energy flow.

Expansion of the Universe and Dark Energy

Key Findings: Observations of distant galaxies reveal that the universe is expanding at an accelerating rate, driven by an unknown force termed dark energy.
The accelerating expansion suggests that the fabric of space-time is dynamically evolving, challenging static views of the universe.

TDM Interpretation: TDM views the expansion of the universe as a result of continuous state activation from the twilight dimension into the reality dimension. As states are activated, energy is injected into the reality dimension, driving the expansion of space. Regions of the universe with less resistance to state activation (e.g., cosmic voids) expand more rapidly, creating variations in the rate of expansion. This process aligns with dark energy as an emergent effect of interdimensional energy flow, linking the progression of time to the large-scale dynamics of the cosmos.

Cosmic Microwave Background (CMB) and Initial Conditions of the Universe

Key Findings: The CMB provides a snapshot of the early universe, showing small fluctuations in temperature that correspond to density variations in the primordial universe.
These fluctuations laid the foundation for the formation of galaxies and large-scale structures.

TDM Interpretation: In TDM, the CMB represents the early stages of sequential state activation, where simpler states (e.g., initial density variations) were activated before more complex states (e.g., galaxies and stars). The uniformity of the CMB, combined with its small fluctuations, reflects the structured yet turbulent energy flow from the twilight dimension during the universe's formative stages. This sequential activation aligns with the emergence of time, where the earliest activated states set the causal framework for all subsequent phenomena.
Unified Perspective in TDM

These experiments and observations, when viewed through TDM, reveal a cohesive picture of how time emerges as an interplay of sequential state activation and interdimensional energy flow. TDM not only provides mechanisms for interpreting these results but also bridges the gaps between quantum mechanics, relativity, and cosmology.

Delayed-Choice and Quantum Eraser: Highlight the adaptability of time and causality.
Time Dilation: Demonstrates resistance-driven variations in state activation.

Gravitational Lensing: Shows how resistance affects light, connecting time to spatial distortions.
Universe Expansion: Links the injection of energy through state activation to the progression of time on a cosmic scale.

CMB: Reflects the structured emergence of time from the earliest stages of state activation.
The experimental evidence supporting TDM's concept of time spans scales from the quantum to the cosmic.

By reinterpreting these findings within the framework of interdimensional energy flow and resistance, TDM provides a unifying explanation for phenomena that traditional theories treat as separate. This integration underscores TDM's potential to offer a deeper understanding of time, causality, and the universe's most fundamental mechanisms

Chapter 6: Wave-Particle Duality and Quantum Phenomena

TDM provides a new interpretation of wave-particle duality, viewing it as a result of interdimensional turbulence in the twilight dimension. In experiments like the double-slit, unobserved particles interact turbulently across multiple potential states, creating interference patterns. Observation aligns energy flows, collapsing these states into particle-like behavior. Weak measurements and hybrid behaviors further illustrate the role of partial activation. This chapter also discusses proposed experiments to test TDM's predictions about environmental influences on quantum coherence.

Wave-Particle Duality and Quantum Phenomena in the Twilight Dimension Model (TDM)

In the TDM, wave-particle duality—the observed ability of particles like electrons and photons to exhibit both wave-like and particle-like behavior—is reinterpreted as a consequence of interdimensional turbulence. This turbulence arises from the dynamic interaction of potential states within the twilight dimension before one is activated into the reality dimension. TDM offers a novel framework for understanding this phenomenon, connecting it to the mechanisms of state activation, energy flow, and the observer's role in resolving turbulence.

Traditional View of Wave-Particle Duality

In quantum mechanics, wave-particle duality is a cornerstone principle. Experiments like the double-slit experiment show that particles such as electrons create interference patterns (wave-like behavior) when unobserved, but act as particles when observed. This duality poses a paradox under classical interpretations, as particles and waves are traditionally considered distinct entities.

Quantum mechanics resolves this paradox mathematically by describing particles as wave functions—mathematical constructs representing probabilities of a particle's location and momentum. However, the wave function collapse, where a particle adopts a definite state upon observation, lacks a clear physical explanation.

TDM's Explanation: Interdimensional Turbulence

TDM provides a causal mechanism for wave-particle duality by introducing the concept of interdimensional turbulence in the twilight dimension. Before a state is activated into the reality dimension, multiple potential states interact dynamically in the twilight dimension. These interactions create turbulence—chaotic flows of energy—manifesting as wave-like interference patterns when unobserved.

Wave Behavior: Interaction of Potential States

In the twilight dimension, all possible states coexist, with energy flowing across the nested spheres of potentiality. When a particle's state is unobserved, the energy flow interacts with multiple potential states simultaneously. This interaction creates patterns akin to waves in a fluid, with peaks and troughs corresponding to the constructive and destructive interference of energy flows. For example, in the double-slit experiment:

As an electron or photon passes through the slits, the energy flow from the twilight dimension interacts with multiple potential states. These interactions generate turbulence, producing an interference pattern on the detection screen, even if only one particle passes through at a time.

Particle Behavior: Resolution of Turbulence

Observation focuses the energy flow, collapsing interdimensional turbulence and selecting a single potential state for activation in the reality dimension. This process aligns the energy flow, resolving the wave-like interference and producing a localized particle-like outcome. In TDM, this corresponds to the collapse of the wave function in traditional quantum mechanics.

Role of the Observer

The observer plays a critical role in resolving turbulence and shaping the outcome of wave-particle duality. In TDM, observation is not merely a passive act but an active interaction with the twilight dimension:

Observation directs the energy flow, suppressing turbulence and ensuring that only one potential state is activated into the reality dimension. This process eliminates interference, transforming wave-like behavior into particle-like behavior, consistent with the observer's reality. Delayed-choice and quantum eraser experiments further validate this view, showing that the timing and nature of observation retroactively influence the turbulence and alignment of potential states, determining the particle's behavior.

TDM's Unified View of Quantum Phenomena

Wave-particle duality is just one of many quantum phenomena that TDM reinterprets through interdimensional turbulence and state activation. Other phenomena include:

Superposition: Reflects the coexistence of multiple un-activated states in the twilight dimension. Turbulence arises from their dynamic interaction, resolved only upon observation.

Entanglement: Indicates shared turbulence between states, where the activation of one state aligns or influences another, even across vast distances.

Wave function Collapse: Represents the resolution of turbulence through focused energy flow, resulting in the activation of a single state.

Experimental Support
Several experiments align with TDM's explanation of wave-particle duality:

Double-Slit Experiment: Demonstrates the wave-like interference of unobserved particles and particle-like behavior upon observation. TDM explains this as turbulence resolving into a single state upon observation.

Delayed-Choice Experiments: Show that the act of observation retroactively influences the particle's behavior, consistent with TDM's view that observation modifies energy flow and turbulence in the twilight dimension.

Weak Measurements: Reveal partial collapse of wavefunctions, producing hybrid wave-particle behaviors. TDM attributes this to partial resolution of turbulence, where some interdimensional flows persist.

Implications for Quantum Mechanics
TDM provides a physical mechanism for phenomena traditionally treated as abstract in quantum mechanics:

Wave-Particle Duality: Results from turbulence and its resolution, rather than an inherent duality in the particle itself.

Observer Effect: Becomes a dynamic interaction with interdimensional flows, shaping the observed reality.

Quantum Uncertainty: Reflects the chaotic nature of interdimensional turbulence before state activation.

In the Twilight Dimension Model, wave-particle duality emerges as a natural consequence of interdimensional turbulence in the twilight dimension. When unobserved, particles exhibit wave-like interference patterns caused by the chaotic interaction of potential states. Observation resolves this turbulence, aligning energy flow to activate a single state, producing particle-like behavior. This reinterpretation not only demystifies quantum phenomena but also provides a unifying framework that connects the quantum world to the broader mechanisms of state activation and energy dynamics in TDM.

The Double-Slit Experiment in the Twilight Dimension Model
The **double-slit experiment** is one of the most famous demonstrations of quantum mechanics, revealing the puzzling nature of **wave-particle duality**. When particles such as photons or electrons pass through two slits, they produce an **interference pattern** on a detection screen, behaving like waves. However, when one slit is observed or measured, the interference pattern disappears, and the particles behave like discrete particles. TheTDM provides a comprehensive explanation for this phenomenon by describing it as the result of **interdimensional turbulence** in the twilight dimension.

Traditional Interpretation of the Double-Slit Experiment

In traditional quantum mechanics, particles are described by a **wave-function**, a mathematical representation of all possible paths or states the particle can take. The wave-function spreads through both slits simultaneously, creating interference patterns. When observed, the wave-function "collapses," and the particle takes a definite path, behaving as a particle rather than a wave.

While this interpretation successfully predicts the outcomes, it does not provide a physical mechanism for why observation causes wave-function collapse or how the interference pattern arises.

TDM's Explanation: Interaction of Potential States

In TDM, the interference pattern is explained as the result of **turbulence** within the twilight dimension, where un-activated potential states interact dynamically. These states exist in the nested spheres of the twilight dimension and are influenced by the flow of energy between the twilight and reality dimensions.

Unobserved Particles: Turbulence Creates Interference

When particles pass through the slits unobserved: The energy flow from the twilight dimension interacts with **multiple potential states** simultaneously, as the particle's exact state is not yet determined. These potential states dynamically interfere with each other, creating turbulence in the twilight dimension. This turbulence manifests in the reality dimension as the interference pattern observed on the screen. The peaks and troughs of the interference pattern correspond to regions of **constructive and destructive interference** in the twilight dimension's energy flows. In essence, the wave-like behavior of the particle reflects the chaotic, turbulent interactions of its un-activated potential states in the twilight dimension.

Observed Particles: Resolving Turbulence

When the particle is observed: Observation focuses the energy flow, collapsing the turbulence in the twilight dimension. This directed energy selects a single potential state, activating it into the reality dimension. As a result, the particle behaves as if it passed through only one slit, eliminating the interference pattern.

The observer's interaction suppresses the chaotic interplay of potential states, creating a localized, particle-like outcome in the reality dimension.

Mechanics of Interdimensional Turbulence

The turbulence that generates the interference pattern arises from the chaotic flow of energy across the nested spheres of the twilight dimension. The interaction of potential states creates regions where energy flows constructively (increasing intensity) or destructively (reducing intensity). This turbulence is translated into the reality dimension as wave-like patterns, observable on the detection screen.

Key factors influencing this turbulence include:

Environmental Conditions: Variations in electromagnetic fields or temperature can alter the turbulence, modulating the interference pattern.

State Complexity: Simpler states closer to the outer spheres produce more defined turbulence, while deeper, more complex states create subtler interactions.

Observation: By focusing the energy flow, observation reduces the chaotic interactions, resolving the turbulence and collapsing the interference into a single outcome.

Delayed-Choice and Quantum Eraser Variants
TDM's explanation extends to delayed-choice and quantum eraser experiments, which demonstrate that decisions about observation made after the particle has passed through the slits can retroactively influence its behavior:

Delayed-Choice Experiments: The observer's interaction with the twilight dimension sends feedback that retroactively aligns potential states, resolving turbulence even after the particle's initial passage.

Quantum Eraser Experiments: Erasing or restoring information about the particle's path modulates the turbulence in the twilight dimension, allowing the interference pattern to reappear or disappear.

These findings support TDM's view that observation actively shapes energy flow between dimensions, rather than merely collapsing a preexisting wave-function.

Experimental Observations Explained by TDM

Interference Pattern: Unobserved particles interact turbulently in the twilight dimension, creating wave-like interference patterns in the reality dimension.

Collapse of Interference: Observation focuses energy flow, suppressing turbulence and activating a single state, resulting in particle-like behavior.

Delayed-Choice Effects: Feedback from the reality dimension influences alignment in the twilight dimension, demonstrating the interconnectedness of dimensions in TDM.
Weak Measurements:

Partial observation creates hybrid behaviors, where some turbulence persists, reflecting incomplete resolution of potential states.

Implications for Quantum Mechanics
TDM reframes wave-particle duality as a dynamic interplay of energy flows and state activation: The wave-like behavior is not an inherent property of the particle but a reflection of interdimensional turbulence. Observation actively resolves this turbulence, shaping the particle's

state in the reality dimension. The model offers a causal explanation for phenomena traditionally treated as probabilistic, connecting quantum mechanics to interdimensional dynamics.

The double-slit experiment, long a hallmark of quantum mechanics, finds a natural explanation within the TDM. The wave-like interference patterns result from the chaotic turbulence of un-activated states in the twilight dimension, while observation focuses energy flow to resolve this turbulence and activate a single state. By linking wave-particle duality to interdimensional dynamics, TDM not only clarifies this quantum phenomenon but also provides a unified framework that bridges the gap between quantum behavior and larger-scale processes.

Weak Measurements, Partial Collapse, and the Role of Observation in the Twilight Dimension Model

In the TDM, the act of observation plays a critical role in shaping the transition of potential states in the twilight dimension into activated states in the reality dimension. While traditional quantum mechanics describes observation as causing the "collapse" of a wave-function, TDM introduces a more nuanced perspective. Observation in TDM is an active interaction that aligns interdimensional energy flows, resolving turbulence in the twilight dimension and determining which state becomes part of observable reality. Weak measurements—experiments that partially observe quantum systems without fully collapsing their wave-like behavior—offer compelling insights into how this interaction operates and validate TDM's predictions about **partial collapse** and **persistent turbulence**.

Understanding Weak Measurements and Partial Collapse

Weak measurements are quantum experiments designed to gather limited information about a system without fully disrupting its superposition of states. These measurements result in outcomes that are **incomplete**, leaving the system in a hybrid state that retains aspects of both its wave-like and particle-like behavior.

In TDM, these outcomes are explained by the concept of **partial turbulence resolution**:
Unobserved Systems: When no observation occurs, energy flows in the twilight dimension interact dynamically across multiple potential states, creating turbulence. This turbulence manifests as wave-like interference patterns in the reality dimension.
Full Observation: A full observation focuses the energy flow, collapsing the turbulence entirely and selecting a single potential state for activation. This results in particle-like behavior, where a single outcome becomes measurable.

Weak Measurements: Partial observation does not fully collapse the turbulence in the twilight dimension. Instead, it reduces the chaotic interactions, allowing some residual turbulence to persist. This creates a **hybrid state**, where the system displays mixed wave-particle characteristics.

Experimental Evidence for Weak Measurements and Partial Collapse

Several experiments highlight the persistence of wave-like properties under weak measurement conditions:

Partial Interference Patterns: When weak measurements are applied, interference patterns do not disappear entirely but are diminished. This supports the idea that some turbulence remains unresolved, allowing wave-like behavior to coexist with particle-like outcomes.

Quantum Superposition in Weak Measurement: Weak measurements have demonstrated that quantum systems can retain aspects of superposition even after partial observation. For example, photons passing through a polarizing filter under weak measurement conditions exhibit partial alignment with the filter while retaining interference patterns.

Hybrid Behaviors: Hybrid outcomes observed in weak measurement experiments align with TDM's prediction that incomplete resolution of turbulence leaves the system in a state where both wave and particle behaviors can be detected simultaneously.

The Role of Observation in Aligning Energy Flows

In TDM, observation is not merely a passive act of measurement but an **active interaction** that modulates the energy flow between the twilight and reality dimensions. This modulation occurs in the following ways:

Focusing Energy Flow: Observation directs the energy flow, suppressing turbulence and aligning the dynamic interactions of potential states in the twilight dimension. This alignment ensures that a specific state is selected and activated in the reality dimension.
In weak measurements, the energy flow is only partially focused, allowing some turbulence to persist, which explains the retention of wave-like properties.
Collapsing States: Full observation collapses the interactions in the twilight dimension into a single, localized state. This collapse eliminates turbulence, producing the particle-like behavior seen in experiments.

TDM extends the idea of collapse by attributing it to the resolution of interdimensional turbulence rather than the wave-function's mathematical collapse.

Feedback to the Twilight Dimension:

Observation also sends feedback into the twilight dimension, influencing the alignment of potential states for future activations. In weak measurements, this feedback is less pronounced, leaving the system in a mixed state where both wave-like and particle-like behaviors coexist.

TDM's Explanation for Wave-Particle Duality in Weak Measurements

Wave-particle duality is traditionally treated as an intrinsic property of quantum systems. TDM reframes this duality as a product of the interaction between the twilight and reality dimensions:
Wave-Like Behavior: Results from unobserved interdimensional turbulence, where multiple potential states dynamically interact, creating interference patterns.

Particle-Like Behavior: Emerges when observation resolves this turbulence, aligning energy flows to activate a single state.

Hybrid Behavior in Weak Measurements: Reflects partial alignment of energy flows, where turbulence is reduced but not entirely resolved. This partial resolution allows wave-like interference to coexist with localized particle outcomes.

For example, in the double-slit experiment with weak measurements: Observing which slit a particle passes through (weakly) does not completely eliminate the interference pattern on the detection screen.

The particle exhibits partial particle-like behavior (slit identification) while retaining wave-like interference patterns (turbulence persistence)

Implications for Quantum Mechanics
The ability of weak measurements to demonstrate hybrid states provides critical validation for TDM's framework:

Partial Collapse: Weak measurements show that observation does not always result in a complete collapse of the wave-function. TDM attributes this to partial resolution of interdimensional turbulence, where some chaotic interactions remain.

Observer Influence: Observation is shown to be a modulating factor that shapes the outcome of state activation. This highlights the observer's role as an active participant in the dynamics of reality, rather than a passive measurer of preexisting states.
Unified Framework:
By incorporating weak measurements, TDM bridges the gap between wave-particle duality and hybrid quantum behaviors, offering a causal explanation for outcomes that traditional quantum mechanics describes only probabilistically.

Weak measurements and partial collapse provide strong support for the Twilight Dimension Model's interpretation of quantum phenomena. The incomplete resolution of turbulence in the twilight dimension explains the hybrid wave-particle behaviors observed in these experiments, while the role of observation is redefined as an active alignment of energy flows that determines the nature of state activation. By integrating weak measurements into its framework, TDM offers a deeper understanding of wave-particle duality, observation, and the mechanisms shaping the quantum world, revealing a cohesive picture of how potential states in the twilight dimension transition into observable reality.

The Role of Observation in TDM: Aligning Energy Flows and Collapsing States
In the TDM, observation is redefined as an active interaction between the observer and the interdimensional energy flows connecting the twilight and reality dimensions. Unlike traditional quantum mechanics, where observation passively "collapses" the wavefunction, TDM posits that observation focuses and aligns these energy flows, resolving the turbulence among potential states in the twilight dimension and selecting a single state for activation in the reality dimension. This process transforms wave-like behavior into particle-like behavior, offering a deeper understanding of quantum phenomena such as wave-particle duality.

Observation as an Energy Flow Modulator

Observation in TDM is not a simple act of measurement; it is a causal interaction that reshapes the energy dynamics between dimensions. Here's how this process unfolds:

Before Observation: Turbulence and Wave-Like Behavior - In the twilight dimension, potential states exist as a nested system of possibilities, dynamically interacting with each other. This interaction creates turbulence in the energy flow, manifesting as wave-like interference patterns in the reality dimension.

For example, in the double-slit experiment, an unobserved particle passes through both slits as a wave of potential states, resulting in an interference pattern caused by the constructive and destructive interactions of the energy flows.

During Observation: Focusing Energy Flow - When an observer measures the system, their interaction directs the energy flow between the dimensions, suppressing turbulence and aligning the potential states toward a specific outcome. Observation acts like a lens, concentrating the chaotic energy into a single pathway.

This alignment forces the selection and activation of one potential state, transforming the system's behavior from wave-like (distributed turbulence) to particle-like (a single localized state).

After Observation: State Collapse and Feedback - The activated state becomes part of the observer's reality dimension, and the turbulence in the twilight dimension is resolved for that event.

Feedback from the act of observation influences the twilight dimension, subtly reshaping the alignment of potential states for future activations. This retroactive alignment explains phenomena like delayed-choice experiments, where observation appears to influence outcomes even after the initial event.

Magnetic Field, Temperature, and Interference Patterns in TDM

TDM predicts that external environmental factors, such as magnetic fields and temperature, can influence the energy flow dynamics and turbulence in the twilight dimension, thereby altering interference patterns. These predictions can be tested experimentally to validate or refine TDM.
Magnetic Fields and Energy Alignment

Magnetic fields influence the behavior of charged particles, such as electrons, and are known to alter quantum states in experiments like the Aharonov–Bohm effect. TDM extends this idea by proposing that magnetic fields can modulate the energy flow between the twilight and reality dimensions, affecting the turbulence responsible for wave-like interference patterns.

Predicted Effects: Weak Magnetic Fields: May shift the alignment of energy flows, creating subtle changes in the interference pattern without fully collapsing the wave-like behavior.

Strong Magnetic Fields: Could significantly reduce turbulence, mimicking the effect of partial observation and leading to hybrid wave-particle behaviors.

Experimental Test: Use a double-slit setup with electrons or photons while applying varying strengths of magnetic fields.
Measure changes in the interference pattern's intensity and structure. A reduction in turbulence (weaker interference) under stronger magnetic fields would support TDM's predictions.

Temperatre and Turbulence
Temperature introduces thermal agitation, which can disrupt quantum coherence and increase noise in quantum systems. TDM predicts that higher temperatures amplify turbulence in the twilight dimension, making it harder for states to align and activate cleanly. This disruption should diminish the clarity of interference patterns, even in the absence of direct observation.

Predicted Effects:
Low Temperatures: Reduce turbulence, allowing for sharper, more defined interference patterns.
High Temperatures: Increase turbulence, leading to diffuse or diminished interference patterns as thermal energy competes with the interdimensional energy flow.

Experimental Test:
Conduct double-slit experiments in environments with precisely controlled temperatures. Compare interference patterns at cryogenic temperatures (minimal turbulence) with those at high temperatures (maximum turbulence). A measurable degradation of interference patterns at higher temperatures would support TDM.

Proposed New Experiments for TDM
To further validate TDM's claims about observation and environmental influences, the following experiments can be designed:

1. Weak Observation with Environmental Modulation
Setup: A weak measurement is applied to a double-slit experiment with simultaneous modulation of magnetic fields and temperature.
Hypothesis: Weak observation should partially resolve turbulence, creating hybrid wave-particle behavior. Environmental modulation (e.g., increased magnetic field strength or temperature) should amplify or suppress the interference patterns.
Expected Outcome: Magnetic fields and temperature modulate the residual turbulence in the twilight dimension, altering the degree of hybrid behavior.

2. Delayed Observation Under Varying Conditions
Setup: A delayed-choice experiment is conducted while varying magnetic fields or temperatures during and after particle passage.
Hypothesis: The timing and environmental conditions of observation influence the alignment of energy flows retroactively.
Expected Outcome: Changes in interference patterns under different conditions demonstrate how external factors and feedback align potential states in the twilight dimension.

3. Magnetic Fields and Quantum Eraser Experiments
Setup: Quantum eraser experiments are conducted with and without applied magnetic fields.
Hypothesis: Magnetic fields interact with the energy flow dynamics, modulating the degree of turbulence in erased or restored interference patterns.
Expected Outcome: Measurable shifts in interference patterns when magnetic fields are applied would indicate their role in influencing turbulence.

Implication for Quantum Mechanics
If these experiments validate TDM's predictions, they would:
Expand the Role of Observation: Demonstrate that observation is not merely a passive act but an active modulator of energy flows, shaping the behavior of quantum systems.
Redefine Environmental Influences: Show that external factors like magnetic fields and temperature directly affect interdimensional turbulence, offering new tools for controlling quantum states.

Bridge Quantum and Classical Physics: Provide a unified explanation for quantum phenomena like wave-particle duality while connecting them to macroscopic factors like temperature and electromagnetic fields.

TDM highlights observation as an interaction that aligns energy flows, resolving turbulence in the twilight dimension and collapsing states into particle-like behavior. This process explains wave-particle duality and hybrid quantum behaviors seen in weak measurements. TDM's predictions about the influence of magnetic fields and temperature on interference patterns open avenues for experimental validation. By testing these effects in controlled setups, researchers can explore the interplay between observation, environmental conditions, and interdimensional energy flows, potentially advancing our understanding of quantum mechanics and the nature of reality itself.

Chapter 7: Large-Scale Applications: Cosmology and Dark Energy

TDM extends its framework to cosmological phenomena, including the accelerated expansion of the universe. It proposes that dark energy is the result of continuous state activation, injecting energy into space and driving expansion. Cosmic voids, as regions of low resistance, expand more rapidly, reflecting the dynamics of state activation. Large-scale structures, such as filaments and voids, are interpreted as manifestations of interdimensional turbulence. This chapter explores how TDM provides a cohesive explanation for the universe's evolution.
Introduction to Cosmological Applications of TDM

The TDM offers a revolutionary perspective by providing a framework that unifies quantum phenomena, cosmological structures, and the evolution of the universe. Traditional physics often struggles to reconcile the principles of quantum mechanics, which govern the microscopic world, with the laws of general relativity, which describe the behavior of large-scale structures such as galaxies and the universe itself. TDM bridges this gap by introducing the concept of the twilight dimension—a static realm containing all potential states—and describing how the flow of energy between this dimension and the reality dimension drives the emergence of time, space, and matter across scales.

At its core, TDM posits that state activation, the process by which potential states in the twilight dimension transition into observable phenomena in the reality dimension, is not limited to quantum particles. This mechanism operates universally, shaping both subatomic particles and vast cosmic structures. On the quantum scale, state activation explains phenomena like wave-particle duality and quantum entanglement. On the cosmological scale, it accounts for the accelerated expansion of the universe, the formation of galaxies, and the intricate patterns of the cosmic web.

Through the lens of TDM, dark energy, often regarded as a mysterious and undefined force, is reinterpreted as the continuous injection of energy into the reality dimension through state activation. This process creates the expansive force driving the accelerated growth of the universe. Similarly, cosmic voids and filaments—the vast empty spaces and dense networks of galaxies that define the large-scale structure of the cosmos—are understood as manifestations of interdimensional turbulence. In regions of low resistance, such as voids, state activation occurs more freely, resulting in faster spatial expansion, while in high-resistance areas like filaments, energy flows concentrate, forming dense structures.

By extending the principles of TDM to cosmology, this framework not only provides a cohesive explanation for phenomena that traditional models treat separately but also opens new pathways for interpreting the evolution of the universe. TDM unites the seemingly disparate behaviors of the microscopic and macroscopic realms under a single paradigm, suggesting that the same interdimensional dynamics governing quantum mechanics also shape the structure and expansion of the cosmos. This chapter will explore how TDM offers an integrated approach to understanding the universe, from its smallest particles to its largest scales.

The Accelerated Expansion of the Universe in TDM

The accelerated expansion of the universe, one of the most puzzling phenomena in modern cosmology, is traditionally attributed to a mysterious force called **dark energy**. While dark energy is well-supported by observational evidence, its nature remains elusive in conventional physics. The Twilight Dimension Model (TDM) offers a compelling explanation by reframing dark energy as the result of **continuous state activation**—the process by which energy flows from the twilight dimension into the reality dimension, bringing potential states into observable existence.

Dark Energy as Continuous State Activation

In TDM, the twilight dimension is a static realm containing all possible states of the universe, arranged in nested spheres of increasing complexity. The reality dimension emerges as these potential states are sequentially activated, driven by energy flow between the dimensions. This process is not uniform or static; it is dynamic and ongoing, constantly introducing new energy into the fabric of space.

Dark energy, under this framework, is not a distinct force or field. Instead, it is the **cumulative effect of the energy injected into the reality dimension during state activation**. Each activation transfers energy from the twilight dimension, subtly increasing the overall energy density of the universe. This energy manifests as an expansive force, causing the universe to grow and accelerate in its expansion.

Energy Injection and the Expansion of Space

The injection of energy during state activation has profound implications for the structure and behavior of the universe. As states are activated, energy flows into the reality dimension, creating a dynamic pressure that drives the expansion of space. This process unfolds in the following ways:

Continuous Expansion: The activation of simpler states, such as fundamental particles and quantum fields, occurred during the early stages of the universe, driving rapid inflation. Today, the activation of more complex states contributes to the ongoing, accelerated expansion of space. The constant influx of energy ensures that the universe continues to expand, even as gravitational forces attempt to pull matter back together.

Acceleration Through Low-Resistance Regions: TDM predicts that regions of low resistance in the twilight dimension, such as cosmic voids, facilitate more efficient state activation. These voids expand more rapidly, contributing disproportionately to the universe's accelerated growth.

This aligns with observational evidence showing that cosmic voids grow faster than denser regions.

Interdimensional Energy Dynamics: The interplay between energy flows and resistance determines the rate of expansion. As the universe evolves, the energy injected through state activation overcomes the gravitational pull of matter, leading to the observed acceleration. This process inherently ties the expansion of space to the mechanics of state activation and interdimensional turbulence.

Evidence Supporting TDM's View of Dark Energy

Observational evidence aligns with TDM's interpretation of dark energy as a product of state activation:

The Cosmic Microwave Background (CMB): Fluctuations in the CMB reflect the early stages of state activation, where energy injected into the universe drove rapid inflation. These patterns provide a snapshot of the energy flows that initiated large-scale structures.

Galaxy Distribution and Cosmic Voids: The accelerated growth of cosmic voids, regions of low matter density, supports TDM's prediction that low-resistance areas facilitate faster state activation and expansion.

The Hubble Tension: Discrepancies in measurements of the universe's expansion rate may reflect variations in state activation dynamics across different regions, as TDM suggests.
The Unification of Dark Energy and State Activation

By linking dark energy to the ongoing process of state activation, TDM unifies the concepts of cosmic expansion and quantum dynamics under a single framework:

Dark energy becomes a natural byproduct of the same interdimensional energy flows that govern phenomena like wave-particle duality and time emergence. The framework provides a mechanistic explanation for why the universe's expansion accelerates over time, grounding cosmological observations in the same principles that explain quantum behavior.
The accelerated expansion of the universe, traditionally attributed to dark energy, finds a cohesive explanation in the Twilight Dimension Model. Through the continuous activation of states from the twilight dimension, energy is injected into the reality dimension, driving the expansion of space. This process not only explains the large-scale growth of the universe but also integrates dark energy into a broader framework that connects quantum phenomena with cosmological evolution. By redefining dark energy as an emergent property of interdimensional energy dynamics, TDM offers a unified perspective on one of the most profound mysteries of the cosmos.

Interdimensional Turbulence and Large-Scale Structures
How turbulence in the twilight dimension influences the distribution of galaxies, clusters, and cosmic web patterns.

Dark Energy: A Dynamic Force of State Activation. Reframing dark energy as an emergent property of energy flow between dimensions.
Evidence from cosmological observations supporting this interpretation.
Cosmic Evolution and Sequential Activation The role of sequential state activation in the formation of galaxies, stars, and large-scale cosmic events.

How TDM explains the emergence of complexity over time.
Testing TDM in Cosmology. Potential experiments and observations to validate TDM's predictions about dark energy and cosmic voids.
Comparing TDM's predictions with data from missions like the James Webb Space Telescope and Euclid.

Unifying Quantum and Cosmological Frameworks

How TDM integrates quantum mechanics and general relativity to explain universal phenomena.
Implications for understanding the interplay between small-scale and large-scale dynamics.
Cosmic Voids and Filaments in the Twilight Dimension Model (TDM)

In the TDM, cosmic voids and filaments are interpreted as manifestations of the energy flow and turbulence dynamics between the twilight dimension and the reality dimension. These large-scale structures, defining features of the cosmic web, are understood not simply as outcomes of gravitational clustering or expansion but as emergent phenomena shaped by the underlying interdimensional processes of state activation.

Cosmic Voids: Regions of Low Resistance
Cosmic voids—vast, under-dense regions of the universe—are traditionally seen as areas where matter is sparse due to gravitational dynamics pulling material toward denser regions. TDM, however, offers a deeper explanation by linking voids to the mechanics of energy flow and resistance during state activation.

Voids as Low-Resistance Zones

Efficient Energy Flow: In TDM, the twilight dimension consists of nested spheres representing all possible states, with each state requiring energy flow for activation. Cosmic voids are interpreted as regions of low resistance in these spheres, where energy flows more freely, facilitating faster state activation.

This efficient activation leads to rapid spatial expansion in these regions, causing the voids to grow larger and more distinct over time.

Minimal Turbulence: The lack of significant matter in voids minimizes interdimensional turbulence. This reduction in chaotic interactions allows for smoother energy flow and faster state transitions, reinforcing the rapid growth and maintenance of voids.

Observable Implications: The accelerated expansion of voids observed in large-scale cosmic surveys aligns with TDM's prediction that low-resistance regions experience more pronounced energy-driven expansion. These regions act as cosmic "pressure release valves," expanding faster than their denser counterparts.

Cosmic Filaments: Concentrated Energy Flows -Cosmic filaments, the dense threads of matter connecting galaxies and clusters, are equally significant in TDM. These structures are reinterpreted as regions of concentrated energy flow shaped by the interplay of resistance and turbulence during state activation.

Filaments as High-Resistance Zones

Localized Energy Concentration: Filaments form in regions where resistance to energy flow is higher, causing energy to accumulate and create dense, elongated structures. This resistance is due to the higher complexity of states being activated, such as galaxies and clusters, which require greater energy input.

As energy is funneled into these high-resistance zones, matter coalesces into the dense patterns observed as filaments.

Role of Turbulence: Interdimensional turbulence in the twilight dimension creates chaotic flows that disrupt smooth energy transfer. In filamentary regions, this turbulence becomes organized, funneling energy along specific pathways.

This organized turbulence aligns with the observed linear or web-like structures of filaments, reflecting the energy dynamics shaping their formation.

Interconnection with Voids: Filaments act as boundaries between cosmic voids, where concentrated energy flow contrasts sharply with the low-resistance regions of voids. This duality emphasizes the dynamic balance of energy flow across the cosmic web.

Cosmic Web Dynamics: A TDM Perspective - The interplay between voids and filaments highlights the broader dynamics of state activation and energy flow in TDM:

Energy Redistribution: Energy flows from low-resistance regions (voids) to high-resistance regions (filaments), creating a self-regulating system where voids expand rapidly while filaments grow denser.

Sequential State Activation: Simpler states are activated more readily in voids, while complex states requiring more energy activation dominate filaments, explaining the hierarchical structure of the cosmic web.

Turbulence as a Structural Force: The turbulence inherent in interdimensional energy flow organizes into large-scale patterns, providing a unifying mechanism for the formation of both voids and filaments.

Supporting Observations and Predictions

TDM's interpretation of cosmic voids and filaments aligns with several key observations:

Growth of Voids:

Observational data from surveys like the Sloan Digital Sky Survey (SDSS) shows that voids expand faster than denser regions. TDM explains this as the result of efficient energy flow and low resistance in these regions.

Filamentary Patterns: The intricate structures of filaments observed in galaxy distributions reflect the organized turbulence and concentrated energy flows predicted by TDM.

Void-Filament Interplay: The stark contrast between voids and filaments mirrors the energy flow dynamics described in TDM, where resistance and turbulence determine the characteristics of each region.

Implications for Cosmology

By redefining voids and filaments as emergent features of interdimensional energy dynamics, TDM provides new insights into the evolution of the universe:
Unified Framework:

TDM unifies the formation of voids and filaments under a single mechanism, connecting quantum energy flow with large-scale cosmic structures.

Dark Energy and Expansion: The accelerated growth of voids contributes to the universe's overall expansion, linking state activation directly to dark energy dynamics.

Future Observations: TDM predicts that variations in void and filament behavior should correlate with differences in interdimensional turbulence and resistance, offering testable hypotheses for future surveys.

Cosmic voids and filaments, fundamental components of the cosmic web, are elegantly explained within the Twilight Dimension Model. Voids emerge as regions of low resistance where energy flows freely, driving rapid expansion, while filaments form as concentrated pathways shaped by turbulence and high resistance. This dynamic interplay not only provides a cohesive understanding of the universe's large-scale structure but also connects quantum principles to cosmological evolution, offering a unified framework for interpreting the cosmos.

TDM as a Cohesive Framework for Interpreting Cosmology

The **Twilight Dimension Model (TDM)** provides a unified perspective on some of the most profound mysteries of the universe, including its **accelerated expansion, large-scale structure,** and the elusive concept of **dark energy**. By integrating interdimensional energy flow, state activation, and turbulence dynamics, TDM bridges quantum mechanics and cosmology, offering an elegant explanation for phenomena that traditional theories treat as distinct or unconnected.

Accelerated Expansion and Dark Energy

In TDM, the accelerated expansion of the universe is a direct result of continuous state activation. The twilight dimension contains all potential states of existence, and energy flowing from this dimension into the reality dimension activates these states sequentially. This constant injection of energy increases the universe's overall energy density, manifesting as a force driving the expansion of space. This process redefines dark energy not as a separate entity but as an emergent property of interdimensional energy dynamics.

Self-Regulating Mechanism: The energy flow is governed by resistance and turbulence in the twilight dimension, ensuring that state activation progresses at a rate that aligns with observed cosmic acceleration.

Void-Driven Expansion: Cosmic voids, regions of low resistance, act as accelerators of expansion, reflecting the dynamics of unimpeded state activation.
By connecting dark energy to the activation process, TDM eliminates the need for hypothetical constructs and places the phenomenon within a broader, unified framework.

Large-Scale Structure: Voids and Filaments

The cosmic web, composed of vast voids and dense filaments, is another feature naturally explained by TDM. These structures emerge from the interplay of resistance, turbulence, and energy flow during state activation:

Voids expand rapidly due to low resistance, representing regions of efficient state activation.
Filaments form in high-resistance zones, where energy accumulates and organizes into dense, elongated patterns.

This interpretation unites quantum turbulence with cosmic-scale structures, demonstrating how interdimensional dynamics influence the organization of matter and space.

Bridging Scales: A Unified Framework

TDM offers a framework that seamlessly connects the microscopic world of quantum mechanics with the macroscopic realm of cosmology:

Wave-Particle Duality: Explained by interdimensional turbulence, the same mechanisms apply to the large-scale distribution of matter in the cosmic web.Emergence of Time: The sequential activation of states in the twilight dimension drives both quantum interactions and the progressive evolution of the universe.

Gravitational Effects: TDM links time dilation, resistance, and state activation to explain gravitational lensing, cosmic void expansion, and the behavior of massive structures.

By integrating these phenomena, TDM unifies concepts traditionally treated as separate domains, offering a cohesive view of reality.

Pathways for Further Research in Cosmology Using TDM

TDM opens new avenues for exploration and experimentation, providing testable predictions and guiding principles for future research:

Investigating Cosmic Voids and Filaments

Prediction: Variations in the growth rates of voids and filaments should correlate with differences in interdimensional turbulence and resistance.

Research: Detailed surveys of void and filament behavior, using tools like the Sloan Digital Sky Survey (SDSS) or the upcoming Euclid mission, can validate these predictions.

Dark Energy as Emergent Energy Flow

Prediction: Regions with faster state activation (e.g., voids) should exhibit higher localized expansion rates, contributing to dark energy's apparent effects.

Research: Measuring differential expansion rates across cosmic structures could provide evidence for state activation dynamics.

Testing Turbulence with Quantum and Cosmological Links

Prediction: Interdimensional turbulence should leave imprints on both quantum interference patterns (e.g., in double-slit experiments) and large-scale cosmic structures.

Research: Comparative studies of turbulence signatures in quantum and cosmic systems could reveal underlying similarities, supporting TDM.

Observing Feedback Mechanisms

Prediction: Feedback from activated states in the reality dimension influences potential states in the twilight dimension, subtly affecting future activations.

Research: Exploring time-dependent phenomena, such as delayed-choice experiments on cosmological scales, may uncover evidence of this feedback.

Implications for Inflationary Models

Prediction: The early rapid expansion of the universe (inflation) was driven by high rates of state activation with minimal resistance.

Research: Analyzing fluctuations in the cosmic microwave background (CMB) can provide insights into the early-state activation dynamics predicted by TDM.

Conclusion: TDM's Contribution to Cosmology

The Twilight Dimension Model redefines our understanding of cosmology by providing a unified framework that connects the quantum and cosmic scales. Through its interpretation of dark energy as continuous state activation, voids and filaments as manifestations of energy flow and resistance, and the interplay of turbulence and activation as the driving force behind large-scale structure, TDM offers a cohesive explanation for the universe's evolution.

By proposing innovative pathways for experimentation, TDM challenges current paradigms and opens the door to a deeper understanding of the cosmos. Future research guided by TDM principles promises to bridge gaps in our knowledge, uniting phenomena across scales and transforming our view of the universe into one of interconnected dynamism and emergent complexity.

Chapter 8: Experimental Validation of TDM

Numerous experiments align with TDM's predictions, including the double-slit experiment, the Aharonov–Bohm effect, and observations of cosmic voids. These studies demonstrate how environmental factors like magnetic fields and temperature influence interference patterns and coherence. Gaps remain, such as the direct measurement of interdimensional energy flows and turbulence. This chapter synthesizes existing experimental evidence, highlights areas for further research, and proposes new experimental setups to validate TDM.

Key Experiments Validating TDM Predictions

The Double-Slit Experiment
The Double-Slit Experiment: Validating the Twilight Dimension Model (TDM)
The double-slit experiment is a classic demonstration of quantum mechanics that reveals the strange behavior of particles like electrons or photons. It shows that particles can behave both like waves, creating interference patterns, and like discrete particles, depending on whether they are observed. This experiment is central to understanding wave-particle duality, and it provides strong indirect evidence for the Twilight Dimension Model (TDM). TDM offers a new interpretation of this phenomenon by introducing the concept of interdimensional turbulence in the twilight dimension.

What Happens in the Double-Slit Experiment?

Imagine a setup where a single particle is fired at a screen with two slits in front of it:

Without Observation: When no one observes which slit the particle passes through, an interference pattern forms on the detection screen. This pattern looks like the waves of water rippling and interfering with each other, suggesting the particle behaves like a wave.

With Observation: When a detector observes which slit the particle passes through, the interference pattern disappears. Instead, the particle behaves like a localized object, hitting the screen in a pattern consistent with it traveling through just one slit.

This behavior creates a puzzle: How can a single particle act like a wave when unobserved and like a particle when observed? How Does TDM Explain This?
TDM introduces the idea of interdimensional turbulence to explain the double-slit experiment. Here's how it works:

The Role of the Twilight Dimension: In TDM, the twilight dimension contains all possible states of a particle, including the paths it could take through each slit.

Before observation, the particle's state in the twilight dimension interacts dynamically with other possible states, creating turbulence. This turbulence is like chaotic waves of energy, where multiple potential paths interfere with one another.

Unobserved Particles: When no observation occurs, the energy flow from the twilight dimension interacts with all possible states simultaneously. This interaction creates the interference pattern seen on the screen, as the turbulence manifests in the reality dimension.

Observed Particles: Observation focuses the energy flow between the twilight and reality dimensions. This process suppresses the turbulence, aligning the energy to activate a single state in the reality dimension.

The result is particle-like behavior: the interference pattern disappears because only one possible path is activated.

TDM describes observation as an active process that reshapes the energy flow, resolving turbulence and collapsing all possibilities into one activated state.

Delayed-Choice Experiments: Strengthening the Case for TDM

A more advanced version of the double-slit experiment is the delayed-choice experiment, where the decision to observe or not is made after the particle has already passed through the slits. Amazingly, the results still change depending on the observation, as though the particle "knew" whether it was being observed—even retroactively.
TDM's Explanation:

The twilight dimension is **timeless**, meaning all potential states coexist without regard to the observer's timeline.

When the observer measures the particle, feedback from the observation influences the alignment of states in the twilight dimension, collapsing turbulence and retroactively aligning the particle's path in the reality dimension. This mechanism explains how the act of observation can appear to affect the past without violating causality.

How Does the Evidence Support TDM?

Wave-Particle Duality: The presence of an interference pattern when unobserved aligns with TDM's description of turbulence in the twilight dimension.
The disappearance of the interference pattern upon observation supports the idea that observation resolves turbulence and activates a single state.

Delayed-Choice Experiments: The retroactive influence of observation fits TDM's concept of a timeless twilight dimension where states remain interconnected and adaptable.

Consistency with Known Physics: TDM's description of turbulence and state activation does not contradict quantum mechanics but adds a deeper, causal explanation for why wave-particle duality occurs.

Why Does This Matter?
The double-slit experiment, especially in its delayed-choice variations, suggests that **observation is not passive**. It actively determines the outcome of reality. TDM expands on this by showing how the interaction between dimensions governs these outcomes:
Without observation, the twilight dimension's turbulence governs the particle's behavior, creating wave-like patterns.

With observation, the flow of energy becomes focused, resolving the turbulence and producing particle-like outcomes.

This framework provides a clearer, more intuitive explanation for the strange quantum phenomena observed in the double-slit experiment, offering a bridge between the unpredictable world of quantum mechanics and a more structured, dimensional model.

The double-slit experiment provides strong evidence for the Twilight Dimension Model. Its results align with TDM's predictions that unobserved particles are influenced by interdimensional turbulence, while observation resolves this turbulence to activate specific states. By offering a causal mechanism for wave-particle duality, TDM not only explains the experiment but also opens the door to a deeper understanding of how observation shapes reality.

The Aharonov–Bohm Effect

The Aharonov–Bohm Effect: Validating the Twilight Dimension Model (TDM)

The Aharonov–Bohm effect is a striking quantum phenomenon that demonstrates how particles, such as electrons, are influenced by magnetic potentials even when they travel through regions where no classical magnetic field is present. This effect has puzzled physicists for decades because it highlights the role of potentials, rather than forces, in determining quantum behavior. The Twilight Dimension Model (TDM) provides a novel interpretation by linking the effect to interdimensional energy flows and their modulation by environmental factors like magnetic potentials.

What Happens in the Aharonov–Bohm Effect?
The Aharonov–Bohm experiment involves sending electrons through a region containing a magnetic potential but no observable magnetic field:

Experimental Setup: A magnetic field is confined within a cylindrical region (e.g., a solenoid), ensuring that the electrons traveling outside this region never encounter the magnetic field directly. Despite this, the electrons passing on either side of the solenoid experience a measurable phase shift in their wave-functions, as observed through interference patterns.

Key Observation: The phase shift occurs purely due to the magnetic potential present in the region, even though there is no classical magnetic force acting on the electrons.
This surprising result challenges the classical view of forces and fields, demonstrating that potentials can influence quantum systems without direct interaction.

How Does TDM Explain the Aharonov–Bohm Effect?
TDM reinterprets the Aharonov–Bohm effect by introducing the concept of interdimensional energy flow and how magnetic potentials modulate turbulence in the twilight dimension. Here's how:

The Role of the Twilight Dimension: In TDM, the twilight dimension contains all possible states of a particle, interacting dynamically before one state is activated into the reality dimension. Magnetic potentials influence the flow of energy between the twilight and reality dimensions, altering the alignment of potential states and introducing phase shifts.

Magnetic Potentials Modulate Energy Flow: The confined magnetic field inside the solenoid creates a magnetic potential outside it, which acts on the energy flow in the twilight dimension. This potential modifies the turbulence in the twilight dimension, subtly shifting the alignment of states.

The altered turbulence translates into a phase shift in the electron's wavefunction when activated in the reality dimension.

No Direct Field Required: Because TDM focuses on energy flow between dimensions, the influence of the magnetic potential does not require the presence of a classical magnetic field in the region where the electrons travel. Instead, the magnetic potential acts as a bridge between dimensions, modulating energy flows and producing observable effects.

Phase Shifts Without Direct Interaction

Observation: Electrons experience a phase shift purely due to the magnetic potential, despite never encountering the magnetic field itself.

TDM Explanation: The magnetic potential modifies the interdimensional energy flows, causing turbulence in the twilight dimension that results in the observed phase shift. This aligns with TDM's claim that environmental factors like magnetic potentials influence state activation.

Influence of Environmental Factors

Observation: The magnitude of the phase shift depends on the strength of the magnetic potential, not the presence of a direct force or field.

TDM Explanation: Magnetic potentials act as external modulating factors, shaping the alignment of potential states in the twilight dimension. The phase shift is a direct consequence of this modulation.

Robustness Across Experimental Conditions

Observation: The Aharonov–Bohm effect persists across a wide range of experimental setups, confirming its fundamental nature.

TDM Explanation: TDM's framework of energy flow and turbulence dynamics provides a universal explanation, connecting the phenomenon to underlying interdimensional processes rather than specific experimental conditions.

How TDM Adds to the Understanding of the Effect: Traditional quantum mechanics explains the Aharonov–Bohm effect by attributing the phase shift to the electron's wave-function interacting with the magnetic potential. While this is mathematically accurate, it lacks a physical mechanism for how potentials influence particles without direct interaction.

TDM provides this mechanism by: Introducing interdimensional energy flow as the medium through which potentials influence particle behavior. Explaining the phase shift as a result of turbulence and alignment changes in the twilight dimension, modulated by the magnetic potential.

This interpretation connects the Aharonov–Bohm effect to a broader framework of environmental influences on interdimensional dynamics, offering a deeper understanding of how quantum systems behave.

Why the Aharonov–Bohm Effect Validates TDM?

Environmental Influence:

The effect directly supports TDM's prediction that external factors like magnetic potentials can modulate interdimensional energy flows and turbulence.

Nonlocality: The ability of magnetic potentials to influence electrons without direct contact aligns with TDM's view of interconnected dimensions and the role of potentials in shaping energy flow.

Phase Shifts: The observed phase shifts correspond to changes in turbulence and alignment predicted by TDM, providing measurable evidence for its interdimensional framework.

The Aharonov–Bohm effect demonstrates that magnetic potentials, even in the absence of classical fields, can influence quantum systems. This aligns with the Twilight Dimension Model's claim that environmental factors modulate interdimensional energy flows, altering turbulence and state alignment. By providing a causal mechanism for the phase shifts observed in the Aharonov–Bohm effect, TDM bridges the gap between abstract quantum principles and physical processes, further validating its framework as a unifying theory of quantum and cosmological phenomena.

Cosmic Voids and Large-Scale Structure

Cosmic Voids and Large-Scale Structure: Validating the Twilight Dimension Model (TDM)
The cosmic web, the intricate large-scale structure of the universe, is characterized by vast cosmic voids and dense filaments connecting galaxies and clusters. These features are often attributed to gravitational dynamics, but the Twilight Dimension Model (TDM) provides a more comprehensive explanation by linking them to the flow of energy between dimensions. According to TDM, voids are regions of low resistance, allowing energy to flow freely and drive rapid expansion, while filaments form as concentrated pathways of energy shaped by interdimensional turbulence.

Cosmic Voids: Regions of Low Resistance - Cosmic voids are enormous, nearly empty regions of space, comprising very little matter compared to surrounding structures. TDM reinterprets voids as areas where state activation in the twilight dimension occurs with minimal resistance, resulting in faster spatial expansion.

How TDM Explains Voids?

Low-Resistance Energy Flow: In the twilight dimension, the activation of potential states requires energy to flow between dimensions. Voids, being regions of low resistance, facilitate this flow more efficiently than denser areas. This free flow leads to a higher rate of state activation, which manifests as accelerated expansion in these regions.

Self-Amplifying Growth: The rapid activation of states in voids reduces the density of matter further, reinforcing their low-resistance nature and causing them to expand faster than surrounding regions. This process aligns with observational data showing that voids grow disproportionately over time.

Observable Effects: The accelerated expansion of voids contributes significantly to the universe's overall growth, highlighting their role as critical drivers of cosmic evolution

Filaments: Concentrated Energy Pathways
Filaments are dense, thread-like structures connecting galaxies and clusters, forming the backbone of the cosmic web. TDM interprets filaments as regions of **high resistance** where energy flow becomes concentrated due to **turbulence in the twilight dimension**.

How TDM Explains Filaments?

Energy Accumulation: In areas of high resistance, energy flow encounters greater turbulence, slowing state activation and causing energy to accumulate along specific pathways. This accumulation forms the dense patterns observed as filaments.

Organized Turbulence: The turbulence in the twilight dimension, while chaotic in voids, becomes organized along filamentary structures. This alignment directs energy flow into linear patterns, shaping the dense threads that connect galaxies and clusters.

Boundary Formation: Filaments form the boundaries of cosmic voids, acting as energy "dams" that separate regions of rapid expansion (voids) from areas of concentrated matter.
Observational Evidence Supporting TDM

Cosmic Voids Expand Faster Than Denser Regions

Observation: Surveys like the Sloan Digital Sky Survey (SDSS) show that voids expand more rapidly than surrounding dense regions, a trend well-documented in large-scale cosmological studies.

TDM Explanation: This rapid expansion matches TDM's prediction that voids, as regions of low resistance, allow for efficient state activation and faster spatial growth.

Filamentary Patterns Reflect Organized Energy Flows

Observation: Filaments exhibit a highly structured, linear distribution, connecting galaxies and clusters in predictable patterns. These structures are consistent across scales and surveys.

TDM Explanation: The dense, organized nature of filaments aligns with TDM's view of turbulence becoming structured in high-resistance regions, directing energy into concentrated pathways.

Void-Filament Interplay

Observation: The boundaries between voids and filaments display sharp contrasts in density and behavior, emphasizing their complementary roles in cosmic evolution.

TDM Explanation: TDM predicts that energy flows freely in voids but becomes turbulent and concentrated in filaments, creating the stark structural differences observed in the cosmic web.
TDM's Unified Explanation for Large-Scale Structure

Traditional models often treat voids and filaments as outcomes of gravitational dynamics alone. While gravity plays a role, TDM offers a deeper, unified explanation:

Energy Flow as a Driving Force: Void expansion and filament formation are governed by the dynamics of interdimensional energy flow, with resistance and turbulence determining their characteristics.

State Activation Across Scales: The same process of state activation that governs quantum phenomena drives the emergence of cosmic structures, linking the microscopic and macroscopic under a single framework.

Turbulence as a Creative Mechanism: The chaotic but directed turbulence in the twilight dimension explains the structured complexity of the cosmic web, providing a causal mechanism for its formation.

Propose Experiments and Observations

Void Expansion Rates

Setup: Use advanced telescopes like the Euclid mission to measure differential expansion rates in voids and compare them to dense regions.
Hypothesis: Voids expand faster due to efficient energy flow and low resistance.
Expected Outcome: Confirmation of accelerated void expansion consistent with TDM predictions.

Filament Energy Distribution

Setup: Analyze energy densities along filaments using galaxy motion and clustering data.
Hypothesis: Filaments concentrate energy flow due to turbulence-induced alignment.
Expected Outcome: Observed energy accumulation and alignment along filaments would support TDM.

Cosmic voids and filaments, the defining features of the cosmic web, find a cohesive explanation within the Twilight Dimension Model. Voids emerge as regions of low resistance, enabling rapid state activation and expansion, while filaments form through concentrated energy flows shaped by turbulence in the twilight dimension. Observational evidence, from void expansion rates to filamentary patterns, strongly aligns with TDM's predictions, offering a unified framework for understanding the universe's large-scale structure and its dynamic evolution. By linking quantum principles to cosmological phenomena, TDM bridges the gap between the smallest and largest scales, providing a transformative perspective on the cosmos.

Gravitational Time Dilation: A TDM Perspective

Gravitational time dilation, a cornerstone of Einstein's general relativity, describes the phenomenon where time passes more slowly in stronger gravitational fields. While general relativity explains this effect as the warping of space-time, the Twilight Dimension Model (TDM) provides an alternative interpretation. TDM posits that time dilation results from **gravitational resistance** within the twilight dimension, which slows the activation of potential states into observable states in the reality dimension. This resistance fundamentally delays the flow of time itself. Evidence from well-established experiments and observations strongly supports this explanation, providing an additional layer of understanding that bridges quantum mechanics and cosmology.

TDM's Explanation for Gravitational Time Dilation

In TDM, time is an emergent property of **sequential state activation**—the process by which potential states in the twilight dimension transition into the reality dimension. Massive objects, such as planets or black holes, create regions of **high gravitational resistance** that impede the energy flow required for state activation. This resistance causes time to slow in these regions because the activation of all potential states, from quantum processes to macroscopic phenomena, is delayed.

The Role of Gravitational Resistance

Slowing State Activation: The more massive an object, the greater the gravitational resistance it exerts. This resistance disrupts the smooth flow of energy from the twilight dimension, delaying the activation of states in the reality dimension.
As a result, all processes—including the ticking of clocks, biological functions, and quantum transitions—proceed more slowly in regions of stronger gravity.

Perception of Time: Observers in regions of weaker gravitational fields, where resistance is lower, experience time as moving faster. Conversely, those in stronger fields perceive time as passing normally, but to external observers, their time appears stretched.

Energy Flow Dynamics: The impeded flow of energy in high-gravity regions aligns with TDM's broader framework of energy dynamics, showing how gravity modulates interdimensional interactions.

Hafele–Keating Atomic Clock Tests

One of the earliest experimental confirmations of time dilation came from the Hafele–Keating experiments in the 1970s. Atomic clocks were flown aboard airplanes and compared to stationary clocks on the ground. The results showed that:
Clocks at higher altitudes (weaker gravity) ran faster.
Clocks closer to Earth's surface (stronger gravity) ran slower.

TDM Interpretation: In regions of weaker gravity at higher altitudes, gravitational resistance is lower, allowing energy flows from the twilight dimension to activate states more efficiently. This results in a faster progression of time.

Near Earth's surface, the stronger gravitational field increases resistance, delaying state activation and slowing the flow of time. This directly supports TDM's prediction that gravitational resistance governs the rate of state activation and, consequently, time.

Black Holes and Extreme Time Dilation

Observations near black holes provide some of the most striking evidence for gravitational time dilation. The movie *Interstellar*, which incorporated real physics, depicted astronauts experiencing hours near a black hole while years passed for observers farther away. This extreme time dilation has been corroborated by studies of gravitational effects near massive objects.

TDM Interpretation: Black holes create regions of immense gravitational resistance, severely impeding energy flow between the twilight and reality dimensions. This resistance delays state activation to an extreme degree, stretching time in the vicinity of the black hole.
Observers far from the black hole, in lower-resistance regions, perceive time as progressing normally. This contrast in state activation rates aligns perfectly with TDM's explanation of gravitational resistance as a primary factor in time dilation.

Time Dilation in GPS Systems

Modern GPS satellites must account for gravitational time dilation to maintain accurate positioning. Clocks on satellites orbiting Earth run faster than those on the ground because they experience weaker gravity.

TDM Interpretation: The weaker gravitational field at the satellites' altitude reduces resistance, allowing states to activate more efficiently and causing time to flow faster.
This operational requirement of GPS systems provides real-world validation of TDM's claim that gravitational resistance directly affects the flow of time.

How TDM Strengthens the Understanding of Time Dilation?
While general relativity attributes time dilation to the curvature of space-time, TDM offers a complementary perspective by identifying the mechanism of resistance behind this phenomenon:

Linking Time and Energy Flow: TDM connects the flow of time to the dynamics of energy between dimensions, offering a physical explanation for why time slows in strong gravitational fields. This explanation integrates the concept of resistance, which is consistent with both quantum mechanics and macroscopic observations.

Bridging Quantum and Relativity: By interpreting time dilation as a delay in state activation, TDM unifies the quantum processes of state transitions with the large-scale effects of gravity. This connection provides a framework that bridges two previously distinct domains of physics.

Testable Predictions: TDM predicts that time dilation should correlate directly with the rate of quantum state activation in different gravitational environments. This offers a pathway for future experiments to further validate the model.

Gravitational time dilation, traditionally explained through space-time curvature, finds deeper context in the Twilight Dimension Model. By attributing time dilation to gravitational resistance and its effect on state activation, TDM provides a unifying framework that connects quantum mechanics and general relativity. Evidence from atomic clock experiments, GPS systems, and black hole observations aligns strongly with TDM's predictions, validating its approach to understanding time as an emergent property of interdimensional energy dynamics. This expanded understanding not only reinforces existing theories but also opens new pathways for exploring the interplay of quantum and cosmological phenomena.

Thermal Effects on Quantum Systems: Validating the Twilight Dimension Model

Quantum systems are uniquely sensitive to their environment, and temperature is one of the most influential factors affecting their behavior. As temperature increases, quantum properties such as coherence, superposition, and entanglement degrade, resulting in quantum decoherence. Traditional quantum mechanics attributes decoherence to environmental interactions that disrupt the delicate quantum states. However, the Twilight Dimension Model (TDM) offers a novel explanation by linking decoherence to amplified turbulence in the twilight dimension caused by thermal agitation. This turbulence disrupts the energy flows necessary for stable quantum behavior, diminishing coherence and observable interference patterns. Experimental evidence strongly supports this perspective, providing insights that unify quantum and environmental dynamics under the TDM framework.

The Role of Temperature in TDM

In the TDM framework, quantum coherence relies on stable energy flows between the twilight dimension, where potential states exist, and the reality dimension, where states become observable. Temperature increases thermal agitation, introducing disturbances in the twilight dimension. This agitation amplifies turbulence in the interdimensional energy flows, disrupting the process of state activation. The resulting instability manifests as decoherence, where quantum systems lose their unique behaviors and begin to exhibit classical properties.

Low Temperatures and Stability: At low temperatures, thermal agitation is minimal, allowing energy to flow smoothly between the dimensions. This stability supports quantum coherence, enabling phenomena such as interference patterns in double-slit experiments and sustained quantum entanglement.

High Temperatures and Disruption: As temperature rises, increased thermal energy introduces chaotic disturbances into the twilight dimension. This turbulence interferes with the smooth activation of potential states into the reality dimension, leading to a breakdown of quantum coherence.

Observable Effects: The breakdown of coherence is evident in diminished interference patterns, faster loss of entanglement, and the transition of quantum systems to classical behavior under high-temperature conditions

Experimental Evidence Supporting TDM

Quantum Decoherence and Temperature
Quantum decoherence experiments provide direct evidence of how temperature affects quantum systems. These experiments show that as temperature increases, quantum systems lose their ability to maintain coherence, transitioning toward classical behavior.

Observation: At high temperatures, quantum systems display faster decoherence, with interference patterns becoming less defined or disappearing entirely.

TDM Interpretation: Higher temperatures amplify turbulence in the twilight dimension, disrupting the energy flows needed to sustain quantum coherence. This aligns with TDM's prediction that thermal effects degrade the interdimensional stability necessary for quantum behavior.

Reduced Interference Patterns

In double-slit experiments, particles such as photons or electrons exhibit interference patterns that are indicative of wave-like behavior. However, at higher temperatures, these patterns diminish.

Observation: Interference patterns become less distinct as temperature increases, indicating a loss of quantum coherence.

TDM Interpretation: Temperature-induced turbulence in the twilight dimension disrupts the particle's ability to interact coherently with its potential states. This turbulence prevents the stable energy flow required for wave-like interference to emerge.

Loss of Quantum Entanglement

Entangled particles are highly sensitive to environmental conditions, including temperature. As temperature rises, the entangled states degrade more rapidly.

Observation: Increased temperature accelerates the decay of entangled states, reducing their quantum correlations.

TDM Interpretation: Thermal agitation amplifies turbulence, disrupting the alignment of interdimensional energy flows that sustain entangled states. This leads to faster decoherence and loss of quantum correlations

How TDM Advances Understanding of Thermal Effects?

While traditional quantum mechanics describes decoherence as a consequence of environmental interactions, TDM adds a deeper layer of explanation by identifying the mechanism behind this disruption.

Energy Flow and Coherence: TDM connects quantum coherence to the stability of energy flows between dimensions. Temperature disrupts these flows by amplifying turbulence in the twilight dimension, providing a causal explanation for the observed effects.

Unifying Quantum and Environmental Factors: By linking thermal agitation to interdimensional dynamics, TDM integrates environmental influences into a broader framework. This provides a cohesive understanding of how external conditions affect quantum systems.

Explaining the Quantum-Classical Transition: TDM explains how increased turbulence causes quantum systems to lose coherence and transition to classical behavior. This bridges the gap between quantum and classical physics, offering a unified perspective on their interplay
The impact of temperature on quantum systems, particularly through decoherence and the loss of coherence, strongly supports the Twilight Dimension Model. TDM's description of temperature-induced turbulence in the twilight dimension provides a clear mechanism for understanding why quantum systems degrade in high-temperature environments. By linking thermal agitation to interdimensional energy flows, TDM unifies quantum behavior with environmental influences, offering a comprehensive framework that bridges traditional quantum mechanics and broader physical principles. Experimental evidence from decoherence studies, reduced interference

patterns, and entanglement decay aligns closely with TDM's predictions, reinforcing its validity and potential for further exploration.

Chapter 9: Implications for Cosmology and Philosophy

The philosophical implications of TDM are profound, redefining concepts like time, space, and causality as emergent properties. This chapter explores how TDM reshapes our understanding of the universe, from quantum scales to cosmic scales. Applications in neuroscience, information theory, and artificial intelligence are also discussed, highlighting the interdisciplinary potential of TDM. The chapter concludes by addressing the model's implications for our perception of reality and the nature of existence.

Implications for Cosmology and Philosophy

The Twilight Dimension Model (TDM) is more than a theoretical framework for understanding quantum phenomena and cosmic structures; it offers a profound reimagining of fundamental concepts such as time, space, and causality. These are not treated as inherent properties of the universe but as emergent features arising from the interaction between the twilight and reality dimensions. This reinterpretation challenges the classical and even quantum views of the universe, opening new avenues of thought across cosmology, philosophy, and interdisciplinary sciences.

Redefining Time, Space, and Causality

In TDM, time and space are emergent properties created by the sequential activation of states from the twilight dimension. This stands in stark contrast to the traditional view of time and space as preexisting frameworks within which all events unfold. According to TDM, the flow of energy from the twilight dimension into the reality dimension determines the progression of time, while the alignment and activation of states define spatial relationships.

Causality, too, is redefined. It no longer follows a strict linear sequence but emerges from the order in which states are activated. This has profound implications for the philosophical understanding of cause and effect, as it suggests that the fabric of reality is fundamentally dynamic, shaped by energy flows rather than fixed principles. Retro-causality, as seen in delayed-choice experiments, further challenges our classical notions, implying that the present can influence the alignment of states in the twilight dimension, which, in turn, affects the past.

Impact on Cosmology

TDM offers a unified framework for interpreting phenomena at both quantum and cosmic scales. In cosmology, the model sheds light on questions surrounding dark energy, the accelerated expansion of the universe, and the large-scale structure of the cosmos. For instance, TDM interprets dark energy as the result of continuous state activation injecting energy into space. This ongoing process drives the expansion of the universe, particularly in regions of low resistance such as cosmic voids. Filaments, which connect galaxies, are seen as pathways of concentrated energy flows shaped by interdimensional turbulence.

This perspective integrates quantum mechanics with cosmological phenomena, addressing gaps in current theories. It also has the potential to inspire new ways of interpreting the cosmic microwave background, the formation of galaxies, and the behavior of black holes. By viewing the universe as a dynamic interplay of dimensions, TDM provides a framework that could unify the seemingly disparate scales of the quantum and the cosmic.

Applications Beyond Physics

While TDM is rooted in physics, its implications extend into other fields, particularly neuroscience, information theory, and artificial intelligence.
Neuroscience

TDM's concept of state activation offers a novel way to think about brain processes. Neural activity could be understood as the activation of potential states, where the flow of energy through neural networks mirrors the dynamics of interdimensional energy flows. This could provide insights into consciousness, memory, and decision-making processes, which may emerge from the interaction of activated and potential neural states.

Neuroscience Through the Lens of TDM: Unlocking New Possibilities

The Twilight Dimension Model (TDM) offers a transformative way of understanding the human brain by interpreting neural activity as the activation of potential states. This perspective suggests that the brain operates similarly to the energy dynamics proposed in TDM, where potential states exist in the twilight dimension and are activated into reality through energy flow. By applying TDM to neuroscience, we could fundamentally reshape how we understand processes like consciousness, memory, and decision-making. This model also suggests new possibilities for brain-machine interfaces, treatments for neurological disorders, and AI-inspired neural systems.

Understanding Consciousness

Consciousness is one of the most profound mysteries in neuroscience. Traditional theories attempt to explain it as an emergent property of neural networks, but TDM provides a deeper framework. In TDM, consciousness could be seen as the continuous interaction between potential states in the twilight dimension and their activation in the brain's reality dimension.

Implications: If consciousness arises from the dynamic activation of potential states, it may not be localized to the brain but exist as a broader interaction between dimensions.
This view opens the door to studying how external factors, such as environmental stimuli or quantum-like effects in neural structures, might influence conscious awareness.
New experimental methods, such as mapping energy flows or interference patterns in neural activity, could be developed to explore the boundaries of conscious states.

Memory and State Activation

Memory could be reimagined within TDM as the selective reactivation of previously activated states. In this framework, memories are not static but exist as potential states in the twilight dimension, which can be re-accessed or modified based on energy flow dynamics.
Implications: This could explain phenomena like memory recall, where specific states are reactivated with precision, and false memories, which could arise from turbulence disrupting the alignment of energy flows.

Future research could focus on enhancing memory recall or erasing traumatic memories by modulating the energy flows that activate specific states. Techniques like neural stimulation or advanced imaging could be refined to observe how memory states transition between potential and activated forms.

Decision-Making as State Selection

In TDM, decision-making is analogous to the selection of a particular state from a range of potential states in the twilight dimension. Neural networks may operate similarly, with competing pathways representing different potential decisions. The chosen pathway is the state that successfully activates into reality, influenced by factors such as energy alignment and environmental input.

Implications: This could lead to new insights into cognitive biases, where certain states are preferentially activated due to prior turbulence or resistance in the energy flow.
Decision-making processes could be enhanced by artificially guiding energy flow, potentially through brain-computer interfaces or targeted neural stimulation.
TDM-inspired algorithms might improve decision-making in AI systems, drawing parallels between biological and artificial neural networks.

Treatments for Neurological Disorders

TDM's concept of interdimensional turbulence and resistance could provide a novel explanation for neurological disorders. For instance, conditions like Alzheimer's disease or epilepsy might involve disruptions in the energy flow required for state activation.

Implications: Therapies could focus on stabilizing energy flows in affected neural regions to restore normal activation patterns. This could involve advanced techniques like low-level electromagnetic stimulation or quantum-inspired therapies targeting neural coherence.
Disorders such as schizophrenia, characterized by distorted perceptions and thoughts, could be reinterpreted as excessive turbulence disrupting state alignment, leading to innovative diagnostic and treatment approaches.

Brain-Machine Interfaces

TDM provides a theoretical foundation for brain-machine interfaces (BMIs) by emphasizing the activation of potential states. BMIs could be designed to interact with the twilight dimension's potential states, enabling seamless communication between humans and machines.
Implications: Future devices could predict and activate states more effectively by integrating with the brain's energy dynamics.

Neural prosthetics could function more efficiently by aligning their activation patterns with the brain's interdimensional flows, improving outcomes for individuals with motor or sensory impairments.

This approach could also lead to the development of immersive virtual reality systems that directly stimulate neural potential states, enhancing the realism of virtual experiences. Applying TDM to neuroscience offers a revolutionary perspective on brain processes, reframing consciousness, memory, and decision-making as outcomes of dynamic state activation. This model opens new avenues for research and innovation, including improved treatments for neurological disorders, enhanced decision-making systems, and advanced brain-machine interfaces. By integrating TDM principles with current neuroscience, we could bridge the gap between quantum and biological systems, transforming our understanding of the brain and its potential. These possibilities not only expand the scope of neuroscience but also pave the way for interdisciplinary advancements in medicine, technology, and artificial intelligence.
Information Theory

TDM's framework aligns closely with the principles of information theory, where data can be seen as the activation of states from a potential set of possibilities. The model suggests that turbulence in the twilight dimension might mirror entropy in information systems, offering a new lens to study noise, information loss, and error correction.

Information Theory and the Twilight Dimension
The principles of information theory focus on the encoding, transmission, and decoding of data. This field is vital for understanding communication systems, data storage, error correction, and even artificial intelligence. The Twilight Dimension Model (TDM) aligns closely with these principles, offering a new perspective on how information exists and evolves. In TDM, data can be interpreted as the activation of states from a potential set of possibilities within the twilight dimension. Turbulence within this dimension could mirror entropy in information systems, providing a novel framework for addressing challenges like noise, information loss, and error correction. If applied, TDM has the potential to revolutionize areas such as computing, communication, cryptography, and machine learning. Below are five realistic examples illustrating the transformative potential of TDM in information theory.

Noise and Signal Clarity

Noise in communication systems refers to random fluctuations that interfere with the transmission of a signal, often leading to errors or degraded data quality. Traditional information theory manages noise using error-detection and correction codes. However, TDM provides a deeper understanding by interpreting noise as a manifestation of turbulence in the twilight dimension.

What Could Change?
Noise reduction could become more precise by identifying and addressing the interdimensional turbulence disrupting data states. This approach could redefine how we design communication systems.

For example, wireless networks could develop adaptive modulation systems that dynamically counteract turbulence, ensuring clearer signals and fewer errors.
In quantum communication, where noise is a critical challenge, TDM might offer strategies for stabilizing quantum states and minimizing decoherence.

Information Loss and Entropy

Entropy, a measure of uncertainty or randomness in information systems, is a key concept in information theory. In TDM, entropy aligns with the concept of turbulence in the twilight dimension, where potential states interact chaotically before becoming activated in the reality dimension. This reinterpretation provides new tools for studying information loss.

What Could Change?
Data storage systems could use TDM principles to minimize entropy by stabilizing the turbulence that leads to information degradation. For instance, hard drives or cloud storage systems might employ interdimensional energy alignment techniques to preserve data integrity. Archiving systems, such as those used for scientific data or historical records, could benefit from enhanced error correction methods derived from TDM's turbulence stabilization mechanisms.

Error Correction and Redundancy

Error correction codes ensure reliable communication by detecting and correcting errors introduced during transmission. These codes often rely on redundancy, which adds extra data to a message. TDM could improve error correction by offering a framework that mirrors turbulence dynamics in the twilight dimension.

What Could Change?
New error correction algorithms could model turbulence and resistance in the twilight dimension, predicting and correcting data loss more efficiently.
For example, satellite communications, where data often encounters interference, could adopt TDM-based error correction to reduce reliance on redundancy, improving transmission speed and reliability.

Cryptography and Secure Communication

Cryptography relies on encoding data in such a way that only authorized parties can decode it. The security of cryptographic systems often depends on mathematical complexity. TDM could enhance cryptographic techniques by leveraging the concept of state activation and interdimensional dynamics.

What Could Change?

New encryption methods could use the unpredictable turbulence of the twilight dimension as a basis for generating highly secure, non-reproducible encryption keys.

Quantum key distribution, which is already exploring the use of quantum mechanics for secure communication, could integrate TDM principles to create even more robust security systems.

Artificial Intelligence and Information Processing

Artificial intelligence systems process vast amounts of data, relying on algorithms to find patterns and make decisions. TDM's framework could offer a new way to model AI systems by aligning their data processing mechanisms with the state activation processes in the twilight dimension.

What Could Change?

Machine learning algorithms could simulate interdimensional energy flows to optimize pattern recognition, decision-making, and adaptability.

For example, AI models used in autonomous vehicles or predictive analytics could achieve higher accuracy by integrating turbulence mitigation strategies inspired by TDM.

Neural networks might become more resilient to "noisy" or incomplete datasets by adopting mechanisms that stabilize interdimensional energy flows

Real-World Impact: Revolutionizing Data Transmission

By redefining noise as turbulence, TDM could lead to the development of adaptive communication systems that self-correct in real-time. Wireless networks, including 5G and beyond, could become more efficient and reliable, transforming how we connect in our increasingly digital world.

Enhancing Data Storage and Retrieval

With entropy reduction techniques inspired by TDM, future data storage systems could achieve near-perfect reliability, ensuring that even critical scientific or historical data is preserved indefinitely.

Securing Digital Communications

TDM-based cryptography could outpace existing quantum encryption methods, providing unmatched security in financial systems, government communications, and personal data protection.

Accelerating AI Development

TDM principles could elevate AI systems to new levels of sophistication, enabling more accurate predictions, better decision-making, and adaptability to complex, dynamic environments. The integration of TDM into information theory has the potential to redefine how we understand and handle data. By interpreting noise, entropy, and error correction through the lens of interdimensional turbulence, TDM offers innovative solutions to longstanding challenges in communication, storage, and cryptography. Furthermore, its principles could revolutionize artificial intelligence, enhancing its ability to process information and adapt to changing conditions. As we continue to explore the intersection of TDM and information theory, the possibilities for real-world applications are both realistic and transformative, promising to reshape how we interact with and rely on information systems in the future.

In artificial intelligence, TDM's ideas could influence how we conceptualize machine learning and decision-making. The process of state activation resembles how AI systems explore potential solutions and refine their responses based on feedback. Incorporating concepts like turbulence and resistance could lead to the development of more adaptive and robust AI systems that mimic the dynamics of natural systems.

Artificial Intelligence and the Twilight Dimension Model
Artificial intelligence (AI) is fundamentally about making machines capable of problem-solving, learning, and adapting to dynamic environments. The **Twilight Dimension Model (TDM)**, with its focus on state activation, turbulence, and resistance, offers a new framework for conceptualizing how AI systems process information and refine their decision-making. By drawing parallels between the process of state activation in TDM and the way AI explores potential solutions, TDM introduces novel strategies for creating more adaptive, resilient, and efficient AI systems. This framework could reshape machine learning, decision-making processes, and system adaptability, offering new paths for innovation and improvement.

Rethinking Decision-Making in AI

AI systems, especially those based on machine learning, operate by exploring a range of possible solutions and selecting the optimal one based on feedback. TDM interprets decision-making as the activation of one state from a set of potential states in the twilight dimension. Turbulence and resistance in this activation process mirror the challenges faced by AI systems when navigating complex datasets or uncertain environments.

What Could Change: AI models could be designed to simulate TDM's state activation process, allowing them to prioritize and activate potential solutions based on environmental factors, similar to how turbulence and resistance influence activation in TDM.
Decision-making algorithms could incorporate TDM-inspired concepts of resistance, enabling AI to weigh solutions more dynamically and adaptively, improving performance in uncertain or rapidly changing conditions.

For example, in autonomous vehicles, algorithms inspired by TDM could better account for dynamic factors such as unexpected obstacles or changing road conditions, leading to safer and more reliable decision-making.

Enhancing Machine Learning Through Turbulence

In TDM, turbulence describes the chaotic interaction of potential states before a single state is activated. This concept aligns with the "exploration" phase of machine learning, where systems navigate through complex datasets to identify patterns and refine their models.

What Could Change: Turbulence could be modeled within machine learning systems to simulate the chaotic dynamics of real-world data, improving their ability to handle noisy or incomplete datasets.

Reinforcement learning algorithms, which rely on trial and error to optimize outcomes, could incorporate TDM's turbulence dynamics to explore more diverse solutions before settling on a final activation.

For example, AI systems used in financial markets could better predict trends by simulating the chaotic interactions of potential economic states, reducing risks and improving investment strategies.

Addressing AI Bias with Resistance Modeling

Bias in AI arises when algorithms disproportionately favor certain outcomes due to imbalanced training data or flawed decision-making processes. TDM introduces the concept of resistance, which impedes state activation, as a mechanism to counterbalance bias.

What Could Change?
By incorporating resistance into AI models, systems could simulate counterforces to biased tendencies, ensuring a more balanced exploration of solutions.
For example, AI systems used in hiring processes could model resistance to amplify underrepresented candidate profiles, reducing systemic biases and promoting fairness.
This approach could also help systems avoid overfitting, where AI becomes too narrowly focused on specific patterns in the training data, by introducing a resistance-based penalty for overly simplistic solutions.

Building Adaptive Systems

Adaptability is a hallmark of intelligent systems, whether biological or artificial. TDM's framework, with its emphasis on dynamic energy flows, provides a blueprint for creating AI systems that can adapt to new environments or challenges by realigning their internal state activation processes.

What Could Change?
AI systems could mimic the interdimensional energy flow described in TDM, allowing them to adapt their activation dynamics in response to new inputs or changing circumstances.
For example, robotics systems in disaster response could use TDM-inspired adaptability to recalibrate their behaviors in unpredictable environments, such as navigating collapsing structures or locating survivors in extreme conditions.
This dynamic approach could also enhance personalization in AI, enabling systems to continuously adapt their responses to individual user preferences or needs.

Improving AI Roustness and Resilience

AI systems often struggle with stability and reliability, particularly when faced with unexpected inputs or adversarial attacks. TDM's turbulence-resistance interplay offers a mechanism for improving system robustness by stabilizing energy flows during state activation.

What Could Change?
Resilience could be built into AI systems by simulating turbulence as a means of stress-testing models before deployment, ensuring they can handle chaotic or adversarial conditions.
For example, AI systems in cybersecurity could be enhanced to withstand hacking attempts by modeling resistance to turbulence, maintaining functionality even under attack.
This could also extend to AI systems used in critical applications, such as healthcare or transportation, where system failure is not an option

Real-World Applications: The potential applications of TDM-inspired AI are vast and transformative. For instance:

Dynamic Algorithms: Search engines and recommendation systems could use turbulence-based exploration to provide more diverse and meaningful results.

Personalized Healthcare: AI systems in medicine could model patient-specific resistance factors to deliver tailored treatment plans.

Sustainable Systems: AI in environmental modeling could simulate turbulence to predict and mitigate the effects of climate change more accurately.

Human-Machine Collaboration: AI interfaces could incorporate TDM principles to align more intuitively with human thought processes, enhancing collaboration in workplaces or education.

Next-Generation Neural Networks: AI architectures could use TDM to develop more human-like learning patterns, bridging the gap between artificial and biological intelligence.
The Twilight Dimension Model offers a revolutionary perspective on artificial intelligence by framing machine learning, decision-making, and adaptability in terms of state activation,

turbulence, and resistance. By integrating TDM's concepts, AI systems could become more robust, adaptive, and capable of handling complex, dynamic challenges. From reducing bias to enhancing resilience and personalization, TDM-inspired AI holds the potential to redefine the field, creating systems that better mimic the intelligence and flexibility of natural systems. As research progresses, the interplay between TDM and AI could lead to breakthroughs that transform industries and improve our ability to navigate an increasingly data-driven world.
Philosophical Implications.

At its core, TDM challenges the very nature of existence and our perception of reality. If time, space, and causality are emergent rather than fundamental, what does this mean for our understanding of the universe? TDM suggests that reality is not fixed but dynamic, shaped by the interplay of potential and observable states. This perspective aligns with certain philosophical traditions, such as process philosophy, which emphasizes becoming over being.
Moreover, TDM forces us to reconsider the role of the observer. In this model, observation is not a passive act but an active process that aligns energy flows and shapes reality. This elevates the role of consciousness and measurement, suggesting that reality is co-created through interaction with the twilight dimension.

Finally, TDM touches on existential questions. If reality is shaped by energy flows and turbulence, does this imply a form of interconnectedness that underlies all phenomena? The model hints at a universe where boundaries between the physical and metaphysical blur, offering a framework for exploring questions about the nature of life, the universe, and our place within it.

The Twilight Dimension Model is more than a scientific theory; it is a paradigm shift that redefines fundamental concepts across disciplines. By presenting time, space, and causality as emergent properties, TDM bridges quantum mechanics and cosmology while opening new pathways for interdisciplinary exploration. Its applications in neuroscience, information theory, and artificial intelligence demonstrate its potential to influence diverse fields, while its philosophical implications challenge us to rethink our perception of reality and existence. As we delve deeper into this model, it promises to transform not only our understanding of the universe but also the ways in which we navigate the mysteries of life and consciousness.

Chapter 10: Future Directions and Open Questions

TDM opens new avenues for research, including experimental validation of interdimensional turbulence and computational modeling of state activation dynamics. This chapter identifies key challenges, such as integrating TDM with existing paradigms and addressing unanswered questions about dark energy and causality. Suggestions for interdisciplinary collaboration are provided, emphasizing the need for innovative approaches to test and refine the model. The chapter concludes with a vision for TDM's role in shaping the future of physics.

The Twilight Dimension Model (TDM) introduces a groundbreaking approach to understanding reality, suggesting that observable phenomena emerge from the interaction between the twilight and reality dimensions through processes such as state activation and interdimensional turbulence. While the model provides a cohesive framework that bridges quantum mechanics, relativity, and cosmology, its acceptance and practical application depend on addressing key challenges and open questions. This chapter outlines realistic advancements that can be made using today's technology, from experimental validation and computational modeling to interdisciplinary collaboration, while also identifying the hurdles that must be overcome for TDM to gain broader acceptance.

A primary avenue for future research involves the experimental validation of interdimensional turbulence, one of TDM's core concepts. Advances in quantum experimentation make this goal feasible. For instance, modifications to the double-slit experiment could shed light on how environmental factors such as magnetic fields and temperature influence interference patterns. By systematically applying these factors during experiments with electrons or photons, researchers can test whether the predicted changes in turbulence align with TDM's claims. Similarly, weak measurement techniques, which allow for partial observation of quantum states without fully collapsing their wave-function, can be employed to explore the residual turbulence in the twilight dimension. If such experiments demonstrate hybrid wave-particle behaviors consistent with TDM, they would lend significant support to its framework.

Computational modeling offers another crucial pathway for advancing TDM. Contemporary computational tools allow researchers to simulate energy flow and state activation dynamics, two fundamental aspects of the model. By adapting techniques from computational fluid dynamics (CFD) and finite element analysis (FEA), scientists can model the flow of energy between the twilight and reality dimensions, capturing how turbulence and resistance shape the activation process. Machine learning algorithms could further enhance these efforts by analyzing experimental data to identify patterns in state activation and turbulence. Such simulations would not only refine the theoretical underpinnings of TDM but also provide valuable insights for designing more targeted experimental setups.

Integrating TDM with existing scientific paradigms is essential for its broader acceptance. This task involves reconciling the model with quantum mechanics, general relativity, and the Standard Model of particle physics. For instance, TDM's explanation of wave-particle duality as a manifestation of interdimensional turbulence could be mathematically formalized and compared with the probabilistic framework of quantum mechanics. Similarly, TDM's reinterpretation of gravity as resistance to energy flow during state activation must be shown to align with Einstein's equations of general relativity while extending them to account for phenomena like dark energy. These efforts would not only validate TDM's claims but also demonstrate its ability to address unresolved questions in physics, such as the incompatibility of quantum mechanics and relativity.

The nature of dark energy and the role of causality represent two of the most profound open questions in TDM. Current astronomical technologies can be leveraged to explore these issues. For example, TDM proposes that dark energy arises from the continuous activation of states in the twilight dimension, injecting energy into the reality dimension and driving the universe's accelerated expansion. By analyzing data from cosmic microwave background radiation and galaxy surveys, researchers can test whether patterns in cosmic void expansion align with TDM's predictions. Similarly, delayed-choice quantum experiments offer a means of investigating TDM's claims about retrocausality—specifically, the idea that observation feedback can reshape potential states even after initial events have occurred. Such experiments could reveal whether TDM provides a more comprehensive explanation for these phenomena than existing theories.

Interdisciplinary collaboration will play a pivotal role in advancing TDM. The model's complexity demands input from experts across diverse fields, including quantum mechanics, computational physics, cosmology, and applied mathematics. Establishing collaborative research centers dedicated to TDM would foster dialogue among these disciplines, enabling the development of innovative methodologies and technologies. Partnerships with technology companies could further accelerate progress by providing access to advanced computational resources and cutting-edge quantum experimentation platforms. Additionally, interdisciplinary educational programs and workshops could train a new generation of researchers equipped to tackle the unique challenges posed by TDM.

Despite its potential, TDM faces significant challenges that must be addressed. One such challenge is the inherent difficulty of experimentally verifying interdimensional interactions, as these processes occur beyond the direct reach of current observational tools. Moreover, integrating TDM into the broader scientific community will require overcoming skepticism and demonstrating its superiority or complementarity to existing frameworks. The development of robust mathematical models and experimental evidence will be critical in achieving this goal. The vision for TDM's future extends beyond theoretical exploration. If its principles are validated, TDM could redefine our understanding of time, space, and reality itself, providing a unified framework that bridges the quantum and cosmic scales. Its applications could revolutionize fields such as quantum computing, cosmology, and materials science, offering new tools for manipulating energy flows and understanding the fundamental nature of existence. In the short term, researchers must focus on achievable milestones, such as validating key predictions through experiments and computational modeling. In the longer term, TDM has the potential to transform physics, bridging gaps between previously disconnected theories and offering profound insights into the fabric of reality.

By addressing its open questions and challenges through interdisciplinary collaboration and technological innovation, TDM can advance from a theoretical construct to a foundational pillar of modern physics. The path forward requires both rigorous scientific inquiry and bold imagination, but the rewards—reshaping our understanding of the universe and our place within it—are well worth the effort.

Conclusion

The Twilight Dimension Model (TDM) offers a revolutionary framework for understanding the universe by unifying quantum mechanics, relativity, and cosmology. This theoretical model reimagines fundamental properties such as time, space, and gravity as emergent phenomena, not intrinsic qualities of the universe. According to TDM, all possible states of reality exist statically in a timeless "twilight dimension." These potential states transition into observable phenomena in the "reality dimension" through energy flows that activate specific states. This interaction between dimensions explains the dynamic nature of the universe and addresses critical gaps left by traditional physics.

One of TDM's most significant contributions is its explanation of time as an emergent property. Time arises from the sequential activation of states, creating the perception of forward progression. Unlike classical physics or relativity, which treat time as either absolute or intertwined with space-time curvature, TDM provides a unified mechanism by linking time's flow to the interdimensional energy transfer. This approach also explains phenomena such as time dilation, attributing the slowing of time near massive objects to increased resistance in energy flow, aligning with relativistic principles but adding a deeper, causal explanation. TDM extends its insights to quantum mechanics, resolving mysteries like wave-particle duality and wave-function collapse. These behaviors are framed as results of interdimensional turbulence—chaotic interactions among un-activated states in the twilight dimension. Observation focuses energy flow, collapsing turbulence and ensuring the activation of a single state, bridging the divide between quantum probability and observed determinism.

On a cosmic scale, TDM provides a novel explanation for dark energy and the universe's accelerated expansion, attributing these phenomena to continuous state activation injecting energy into the reality dimension. By offering a cohesive framework that links quantum mechanics, relativity, and cosmology, TDM opens pathways for transformative research and technological advancements, profoundly redefining our understanding of reality.

Chapter 11: Provisional Patent Applications Examples

Provisional Patent Application: Twilight Dimension Model-Based Neural State Activation System

Title: Twilight Dimension Neural Activation Framework for Enhanced Understanding and Treatment of Neurological Disorders

---

Introduction

The Twilight Dimension Model (TDM) offers a revolutionary perspective on understanding the human brain by analogizing neural activity to the activation of potential states. In this framework, potential states exist in the twilight dimension and are activated into observable reality via directed energy flows. This paradigm challenges conventional neuroscience by suggesting that consciousness, memory, and decision-making arise not solely from biochemical processes but also from dynamic energy interactions across dimensions. TDM can fundamentally reshape brain-machine interfaces, the treatment of neurological disorders, and AI neural networks, fostering advancements in healthcare and technology.

Today's Problem and Its Solution

Problem: Current neuroscience is limited in explaining consciousness, memory retention, and neuroplasticity in a unified manner. Brain-machine interfaces (BMIs) and treatments for conditions such as Alzheimer's, Parkinson's, and epilepsy often rely on partial mechanistic models, leading to suboptimal interventions. Similarly, AI neural networks, though inspired by biological neurons, lack the depth to emulate human cognition's adaptability and complexity.

Solution: By integrating the TDM perspective, we propose a system where neural activation corresponds to state transitions in the twilight dimension. This model not only provides a unified explanation for cognitive phenomena but also enables innovative interventions. For example, it allows the design of energy-modulating devices to enhance state activation, improving memory retrieval or mitigating neurodegenerative impacts. Furthermore, TDM-based algorithms for AI can revolutionize neural systems by replicating the dynamic adaptability of human cognition.

Detailed Description

1. **Framework Overview** The TDM-based system interprets the human brain as a conduit for energy flow between the twilight and reality dimensions. Neural activation is modeled as a state selection process, where latent potential states in the twilight dimension are activated into reality, corresponding to neuronal firing patterns. This dynamic interaction creates the emergent properties of consciousness, memory, and decision-making.

2. **Neural Energy Flow and State Activation**
   - **Potential States:** Each neuron represents a potential state in the twilight dimension. Synaptic connections are modeled as pathways for energy flow, analogous to Tesla valves, guiding the sequence and intensity of state activation.
   - **Activation Dynamics:** Neural firing occurs when energy overcomes resistance within the twilight dimension, influenced by observation (external stimuli) and environmental factors such as magnetic fields or temperature.
3. **Applications in Neurological Treatments**
   - **Memory Retrieval Enhancement:** Devices using targeted electromagnetic fields can reduce turbulence in state activation, facilitating seamless energy flow and improving memory recall.
   - **Neuroplasticity Stimulation:** By modulating resistance in energy flows, the framework allows for the targeted stimulation of synaptic reconfiguration, aiding recovery from trauma or stroke.
   - **Epilepsy Intervention:** Predictive algorithms identify turbulence patterns preceding seizures, enabling real-time modulation of energy flows to stabilize neural activation.
4. **Integration with Brain-Machine Interfaces**
   - BMIs leveraging TDM principles can establish more precise connections between neural activity and external devices. By modeling energy flow as dynamic state activation, BMIs can predict and replicate complex thought patterns with unprecedented accuracy.
5. **Advancements in AI Neural Networks**
   - **State-Based Learning Algorithms:** AI systems inspired by TDM implement turbulence-resistant pathways, enabling adaptive decision-making akin to human cognition.
   - **Multi-Dimensional Analysis:** Neural networks emulate interdimensional energy flows, incorporating latent states for improved problem-solving capabilities.
6. **Prototype Implementation**
   - A wearable headset utilizes low-frequency electromagnetic fields to modulate state activation in the twilight dimension. Initial tests demonstrate enhanced focus, improved cognitive function, and reduced symptoms of neurological disorders.

Claims

1. **Neural State Activation Framework:** A method to model neural activity using the Twilight Dimension Model, wherein each neural activation corresponds to energy-driven state selection within the twilight dimension, enabling enhanced understanding and treatment of neurological conditions.
2. **Energy Modulation Device:** A device for improving cognitive function and treating neurological disorders by modulating energy flow between potential states in the twilight dimension and observable neural activity, using controlled electromagnetic fields.

3. **AI Neural System Inspired by TDM:** An AI framework that integrates turbulence-resistant pathways and state activation principles, allowing for adaptive decision-making and multi-dimensional problem-solving capabilities.

Provisional Patent Application: Dynamic Consciousness Activation Framework

Introduction

The proposed framework redefines the understanding of consciousness by integrating the Twilight Dimension Model (TDM). TDM suggests that consciousness emerges not as a localized phenomenon in the brain but as a dynamic interaction between dimensions, where latent potential states in a timeless twilight dimension are activated into observable realities through energy flows. By applying this model, consciousness is envisioned as an emergent property shaped by external stimuli, quantum-like effects, and neural structures interacting with interdimensional energy patterns. This theory offers novel insights into consciousness, enabling new methodologies for studying its boundaries, developing experimental tools, and pioneering interventions in cognitive science, brain-machine interfaces, and AI neural systems.

Today's Problem and Its Solution

Problem: Despite advancements in neuroscience, the underlying mechanisms of consciousness, memory, and decision-making remain largely speculative. Current frameworks focus on localized brain activities, which fail to account for quantum-scale phenomena or how external environmental factors influence cognition. As a result, existing brain-machine interfaces and treatments for neurological disorders remain limited in their efficacy, and AI neural systems lack the adaptability and complexity to mirror human cognition. Moreover, methods to study or expand conscious states are constrained by technological and theoretical gaps.

Solution: Utilizing TDM, this invention proposes that consciousness arises from the dynamic activation of potential states, extending beyond the brain's physical limits. By developing experimental methods to map energy flows, interference patterns, and environmental effects in neural activity, this system facilitates a deeper understanding of conscious awareness. This model supports the creation of advanced devices and algorithms to manipulate these dynamics, offering transformative possibilities for treating neurological conditions, enhancing brain-machine interfaces, and building adaptive AI systems capable of simulating human-like cognition.

Detailed Description

1. **Conceptual Framework**
    - **Twilight Dimension and Consciousness:** The twilight dimension is a static realm encompassing all potential states of existence. Consciousness arises as these states are dynamically activated into observable reality through interdimensional energy flows, which are influenced by environmental factors, neural structures, and quantum phenomena.

- **Energy Flow Mechanism:** Neural activity corresponds to the activation of potential states, with energy flows modulated by external stimuli and observer influences. These flows encounter resistance, analogous to a Tesla valve, creating turbulence that shapes the nature and sequence of state activation.

2. **Mechanisms of State Activation**
   - **Interference Patterns in Neural Activity:** Neural structures act as conduits for energy flows, where turbulence and interference patterns arise from the interaction of unactivated potential states. These patterns are modulated by external conditions, such as magnetic fields, temperature, and environmental stimuli, which influence conscious perception and decision-making.
   - **Observer Influence:** Observation collapses interdimensional turbulence, aligning energy flows and selecting specific states for activation. This resolves potential states into conscious awareness, linking perception to quantum mechanics.

3. **Applications**
   - **Neurological Treatments:** Devices that apply targeted electromagnetic fields to modulate energy flows can enhance memory, restore cognitive function, and treat disorders like Alzheimer's, Parkinson's, and epilepsy.
   - **Brain-Machine Interfaces (BMIs):** By mapping and manipulating energy flows, BMIs can establish more precise connections between neural activity and external devices, enabling real-time translation of thought into action.
   - **AI Neural Systems:** Inspired by the dynamic adaptability of state activation, AI systems can emulate human decision-making by integrating turbulence-resistant pathways and multi-dimensional analysis.

4. **Experimental Methods**
   - **Mapping Energy Flows:** Techniques using advanced imaging tools can visualize interference patterns in neural activity, offering insights into the dynamics of conscious state activation.
   - **Environmental Modulation:** Controlled experiments involving magnetic fields and temperature variations can assess their impact on energy flow and turbulence, refining models of consciousness and developing targeted interventions.

5. **Prototype Implementation**
   - **Wearable Devices:** A prototype headset applies low-frequency electromagnetic fields to modulate turbulence, enhancing focus, reducing cognitive decline, and exploring altered states of consciousness.

Claims

1. **Dynamic Consciousness Activation System:**
   - A system for modeling and influencing consciousness through the activation of potential states in the twilight dimension, where external stimuli and neural activity modulate energy flows to resolve turbulence into conscious states. This system facilitates the study and enhancement of awareness by mapping energy flows and interference patterns.
2. **Energy Flow Modulation Device:**

- A device that applies targeted electromagnetic fields to neural structures, modulating energy flow dynamics to influence conscious states. This device is designed for enhancing memory retrieval, improving cognitive function, and treating neurological disorders by optimizing state activation processes.
3. **AI Neural Systems Inspired by TDM:**
    - An AI framework that incorporates turbulence-resistant pathways and interdimensional energy dynamics, enabling adaptive decision-making and multi-dimensional problem-solving. This framework replicates human-like cognition by integrating dynamic state activation mechanisms into neural algorithms.

Provisional Patent Application: Neural Energy Flow Modulation System for Advanced Diagnostic and Therapeutic Applications

Introduction

This invention presents a groundbreaking system for diagnosing and treating neurological and cognitive disorders by stabilizing energy flows and managing turbulence in neural regions based on the Twilight Dimension Model (TDM). TDM proposes that neural activity and consciousness are driven by the activation of potential states through energy flow dynamics between dimensions. Disruptions in these flows, such as turbulence or resistance, can lead to conditions like schizophrenia or memory impairments. The proposed system employs techniques like low-level electromagnetic stimulation and quantum-inspired therapies to realign energy flows, reduce turbulence, and restore coherent state activation patterns. This approach redefines how neurological disorders are understood and treated, leveraging insights into interdimensional energy flow dynamics.

Today's Problem and Its Solution

Problem:
Current neuroscience approaches lack a unified framework for understanding complex neurological conditions like schizophrenia, Alzheimer's disease, or traumatic brain injuries. Conventional therapies often treat symptoms without addressing the root causes, such as disrupted neural coherence or misaligned state activation. Disorders characterized by distorted perceptions, such as schizophrenia, remain difficult to diagnose and treat effectively. Furthermore, existing diagnostic tools fail to capture the dynamics of energy flow or turbulence in neural structures, which are key contributors to these disorders.

Solution:
The proposed system introduces a novel diagnostic and therapeutic framework based on TDM principles. It interprets disorders as manifestations of excessive turbulence or misaligned energy flows in neural structures. By using advanced techniques like electromagnetic field modulation and quantum-inspired therapies, the system realigns energy flows, reduces turbulence, and stabilizes state activation patterns. This approach provides a comprehensive method for restoring neural coherence, enabling earlier diagnosis, more precise interventions, and improved outcomes for patients with neurological disorders.

Detailed Description

1. **Conceptual Basis**
    - **Energy Flow Dynamics:** Neural activity is modeled as the activation of potential states through energy flows. These flows are influenced by environmental factors,

neural structures, and interdimensional turbulence. Disruptions in flow dynamics, such as excessive turbulence or resistance, lead to misaligned state activation, causing distorted perceptions or cognitive impairments.
    - **Turbulence as a Diagnostic Marker:** Disorders like schizophrenia are reinterpreted as excessive turbulence disrupting state alignment. This turbulence prevents coherent activation of neural states, leading to symptoms such as hallucinations and cognitive fragmentation.
2. **Diagnostic Approach**
    - **Energy Flow Mapping:** Advanced imaging techniques, such as functional MRI combined with electromagnetic sensors, are used to map turbulence and energy flow patterns in the brain. These maps highlight regions with excessive turbulence or misaligned flows, serving as diagnostic markers for conditions like schizophrenia or traumatic brain injuries.
    - **Neural Coherence Analysis:** Algorithms analyze coherence across neural networks, identifying areas where turbulence disrupts synchronized activity. This data provides actionable insights for targeting therapeutic interventions.
3. **Therapeutic Techniques**
    - **Electromagnetic Stimulation:** Low-level electromagnetic fields are applied to neural regions with disrupted energy flows. By modulating these fields, turbulence is reduced, and coherent state activation is restored. For example, targeted stimulation in the prefrontal cortex can alleviate symptoms of schizophrenia by stabilizing neural coherence.
    - **Quantum-Inspired Therapies:** These therapies involve manipulating quantum-like interference patterns in neural activity to optimize energy flow. Techniques such as low-temperature magnetic resonance or phase coherence modulation are used to realign energy flows, reduce resistance, and enhance cognitive function.
4. **Implementation**
    - **Wearable Therapeutic Device:** A portable device combines sensors for real-time turbulence mapping with electromagnetic field generators. Patients wear the device during therapy sessions, which dynamically adjusts field strengths based on turbulence measurements.
    - **AI-Assisted Intervention:** AI algorithms analyze patient data, predict turbulence patterns, and recommend personalized stimulation protocols for maximum therapeutic efficacy.
5. **Applications**
    - **Schizophrenia Treatment:** The system reduces turbulence in neural regions associated with distorted perception, such as the prefrontal cortex and temporal lobes. This stabilizes state activation, alleviating symptoms like hallucinations and disorganized thought patterns.
    - **Cognitive Rehabilitation:** For conditions like Alzheimer's or traumatic brain injuries, the system enhances neural coherence and memory recall by realigning energy flows in affected regions.
    - **Preventive Diagnostics:** Early-stage turbulence mapping can identify individuals at risk for neurological disorders, enabling timely interventions to prevent disease progression.

Claims

1. **Neural Energy Flow Modulation System:**
   - A diagnostic and therapeutic system that maps and modulates energy flows in neural structures to stabilize state activation patterns. The system includes sensors for real-time turbulence mapping, electromagnetic field generators for reducing turbulence, and AI-assisted algorithms for personalized intervention protocols.
2. **Electromagnetic Field Modulation Device:**
   - A wearable device designed to apply low-level electromagnetic fields to neural regions, dynamically adjusting field strengths based on real-time turbulence measurements. The device restores neural coherence, reduces turbulence, and alleviates symptoms of disorders like schizophrenia, Alzheimer's, and traumatic brain injuries.
3. **Quantum-Inspired Neural Therapy Framework:**
   - A therapeutic method leveraging quantum-like principles to manipulate interference patterns in neural activity. This framework includes techniques such as phase coherence modulation and low-temperature magnetic resonance to realign energy flows, reduce resistance, and optimize cognitive function.

Provisional Patent Application: Twilight Dimension-Based Error Correction System

Introduction

This invention introduces a transformative approach to error correction in communication systems by applying the principles of the Twilight Dimension Model (TDM). Traditional error correction methods rely on redundancy, adding extra data to ensure reliable communication. While effective, these methods increase data overhead and reduce transmission efficiency, particularly in high-interference environments such as satellite communications. The proposed system leverages TDM's framework, which models turbulence and resistance in the twilight dimension, to predict and correct errors with minimal redundancy. By simulating turbulence dynamics, this method enhances the accuracy and efficiency of error correction, reducing data overhead and improving reliability across various communication channels.

Today's Problem and Its Solution

Problem:
Modern communication systems often encounter interference, noise, and data loss during transmission, particularly in high-interference environments like satellite communications, IoT networks, and long-distance wireless systems. Existing error correction codes (e.g., Hamming, Reed-Solomon) mitigate these issues by introducing redundancy, which increases data payload and slows transmission speeds. This approach is inefficient for bandwidth-constrained systems and high-latency applications, where minimizing data overhead is critical. Moreover, conventional error correction techniques struggle to adapt dynamically to unpredictable interference patterns.

Solution:
The proposed TDM-based error correction system reimagines error detection and correction as the resolution of turbulence in the twilight dimension. Instead of relying on redundancy, this method uses algorithms that model data loss and interference as turbulence patterns, dynamically predicting and correcting errors. By applying turbulence dynamics and resistance principles, the system enables real-time, efficient error correction without significantly increasing data payload. This innovation improves transmission speed, reduces redundancy, and enhances reliability in challenging communication environments such as satellite networks and IoT systems.

Detailed Description

1. **Conceptual Framework**
    - **Turbulence in the Twilight Dimension:** The twilight dimension is conceptualized as a realm where potential states of data exist. Data transmission errors occur when turbulence disrupts the transition of these potential states into

observable reality. Modeling this turbulence allows for dynamic prediction and correction of errors.
    - **Energy Flow Dynamics:** Interference and noise are treated as resistive forces that distort energy flows in the twilight dimension. Correcting errors involves stabilizing these flows, reducing turbulence, and realigning data states.
2. **System Components**
    - **TDM-Based Algorithms:** Algorithms model turbulence and resistance in the twilight dimension to predict error patterns and correct data loss dynamically. These models are continuously refined using machine learning, which adapts to specific communication environments.
    - **Interference Pattern Mapping:** Sensors and analyzers map interference patterns in real time, identifying regions of high turbulence. These patterns are translated into corrective actions using the TDM framework.
    - **Error Correction Engine:** A hardware or software module applies predictive corrections based on turbulence analysis, ensuring data integrity with minimal redundancy.
3. **Applications**
    - **Satellite Communications:** By reducing redundancy, the system improves transmission speed and reliability in high-interference environments. For example, TDM-based error correction can adapt dynamically to signal distortions caused by atmospheric conditions or solar activity.
    - **IoT Networks:** The system enhances error correction in bandwidth-constrained IoT networks, enabling reliable communication between devices with limited processing power and storage capacity.
    - **High-Speed Data Transmission:** In fiber-optic and wireless networks, the system mitigates data loss caused by signal attenuation or environmental interference, ensuring high-speed, low-latency communication.
4. **Implementation**
    - **Simulation and Training:** TDM-based algorithms are trained on interference data to simulate turbulence dynamics in specific communication environments. This training enhances the system's predictive accuracy.
    - **Integration with Existing Protocols:** The error correction engine is compatible with standard communication protocols (e.g., TCP/IP, 5G), allowing seamless integration into existing infrastructure.
    - **Prototype Development:** A prototype module integrates TDM-based error correction with satellite communication systems, demonstrating a significant reduction in data loss and redundancy compared to traditional methods.
5. **Advantages**
    - **Reduced Redundancy:** Unlike traditional error correction methods, the TDM-based system minimizes data overhead, increasing transmission efficiency.
    - **Dynamic Adaptation:** The system adapts to real-time interference patterns, ensuring reliable communication in unpredictable environments.
    - **Scalability:** The system is scalable across various communication channels, from satellite networks to IoT systems, without requiring significant infrastructure changes.

Claims

1. **Twilight Dimension-Based Error Correction Framework:**
   - A system for error correction in communication channels, leveraging the Twilight Dimension Model to model data transmission errors as turbulence patterns. The framework predicts and corrects errors dynamically by stabilizing energy flows and resolving turbulence without relying on traditional redundancy-based methods.
2. **Predictive Error Correction Algorithm:**
   - An algorithm that models interference and noise as turbulence in the twilight dimension, identifying and correcting errors by dynamically realigning data states. The algorithm adapts to real-time interference patterns using machine learning, ensuring efficient and reliable data transmission.
3. **Integrated Error Correction Module:**
   - A hardware or software module designed to integrate TDM-based error correction into existing communication systems. The module includes sensors for interference pattern mapping, a turbulence analysis engine, and a predictive correction processor, ensuring minimal data loss and improved transmission speeds across various environments.

Provisional Patent Application: TDM-Based Cryptographic System for Enhanced Data Security

Introduction

This invention introduces a novel cryptographic system leveraging the Twilight Dimension Model (TDM) to generate secure encryption keys based on the unpredictable turbulence and resistance dynamics of interdimensional energy flows. Traditional cryptographic methods rely heavily on mathematical complexity to ensure security, making them vulnerable to advancements in computational power and quantum computing. By incorporating TDM principles, the proposed system creates encryption keys derived from turbulence patterns, which are inherently non-reproducible and resistant to computational attacks. This approach ensures superior security, particularly in applications requiring highly secure communication, such as satellite networks, financial systems, and military operations.

Today's Problem and Its Solution

Problem:
Conventional cryptographic systems depend on mathematical algorithms that, while secure under current computational capabilities, face vulnerabilities as quantum computing and advanced decryption techniques evolve. These methods often require significant computational resources for key generation and encryption, leading to inefficiencies in speed and energy consumption. Additionally, static key generation methods are susceptible to replication or prediction, increasing the risk of data breaches in high-stakes environments like satellite communication and financial transactions.

Solution:
The TDM-based cryptographic system addresses these challenges by using interdimensional turbulence as the foundation for generating encryption keys. This method leverages the unpredictable nature of turbulence in the twilight dimension to create dynamic, non-reproducible keys. The system dynamically maps turbulence patterns during state activation processes, translating these patterns into secure cryptographic keys. Unlike traditional methods, this approach eliminates the need for extensive redundancy, reduces computational overhead, and provides enhanced resistance to brute force and quantum-based decryption attempts.

Detailed Description

1. **Conceptual Framework**
    - **Turbulence and Resistance in TDM:** TDM describes the twilight dimension as a realm of potential states where turbulence arises from interdimensional energy flows. These turbulent flows are unique and non-reproducible, providing a robust basis for encryption key generation.

- **Dynamic Key Generation:** Encryption keys are derived from the real-time mapping of turbulence patterns, ensuring each key is unique and non-static. The inherent unpredictability of turbulence eliminates the risk of key duplication or reverse engineering.

2. **System Components**
   - **Turbulence Sensor Array:** Advanced sensors measure turbulence patterns in real-time during state activation processes. These patterns are captured as high-resolution datasets representing interdimensional energy flows.
   - **Key Generation Module:** This module translates turbulence data into cryptographic keys. Algorithms analyze the turbulence dataset to extract unique features, which are encoded as encryption keys.
   - **Adaptive Encryption Engine:** Integrates TDM-based keys into existing cryptographic protocols, ensuring compatibility with standard encryption frameworks like AES and RSA while enhancing their security.

3. **Operational Workflow**
   - **Initialization:** The system initializes by activating a set of states in the twilight dimension, generating interdimensional turbulence.
   - **Data Capture:** The turbulence sensor array records turbulence patterns as high-resolution datasets, capturing their chaotic and non-linear characteristics.
   - **Key Processing:** The key generation module processes the datasets using TDM-based algorithms to create unique, non-reproducible encryption keys.
   - **Encryption and Transmission:** The adaptive encryption engine uses these keys to encrypt data, which is then transmitted securely over the desired communication channel.
   - **Decryption:** Authorized receivers use synchronized turbulence mapping to recreate the corresponding key, enabling decryption.

4. **Applications**
   - **Satellite Communications:** The system ensures secure data transmission by dynamically generating keys resistant to interference and quantum decryption.
   - **Financial Systems:** Provides secure encryption for high-value transactions, protecting against advanced hacking attempts.
   - **Military Operations:** Enhances the security of classified communications, ensuring data integrity in hostile environments.

5. **Advantages**
   - **Quantum Resistance:** The unpredictable nature of turbulence patterns ensures resistance to quantum decryption techniques.
   - **Dynamic Security:** Keys are generated in real-time and are unique to each transaction, reducing vulnerability to interception or replication.
   - **Efficiency:** Eliminates the need for extensive computational resources, enabling faster and more energy-efficient encryption processes.

Claims

1. **Turbulence-Based Key Generation System:**

- A cryptographic system that leverages turbulence patterns in the twilight dimension to generate unique, non-reproducible encryption keys. The system includes turbulence sensors, a key generation module, and an adaptive encryption engine, providing enhanced security for data transmission and storage.

2. **Dynamic Encryption Key Algorithm:**
    - An algorithm that translates real-time turbulence patterns into secure cryptographic keys. This algorithm ensures each key is unique and resistant to computational and quantum-based decryption techniques by utilizing the non-linear characteristics of interdimensional turbulence.

3. **Integrated Cryptographic Framework:**
    - A framework integrating TDM-based encryption keys with existing cryptographic protocols, ensuring compatibility and enhanced security. The framework includes synchronization modules for real-time key recreation, enabling seamless decryption by authorized receivers.

Provisional Patent Application: Adaptive Machine Learning and Communication System Based on Twilight Dimension Dynamics

Introduction

This invention introduces a groundbreaking approach to optimizing machine learning algorithms and communication systems by leveraging the Twilight Dimension Model (TDM). TDM provides a framework for simulating interdimensional energy flows, treating noise and interference as turbulence. By modeling these dynamics, the proposed system enhances pattern recognition, decision-making, and adaptability in AI and communication systems. This technology allows neural networks to stabilize turbulent energy flows, improving their resilience to noisy or incomplete datasets. Furthermore, TDM-inspired adaptive communication systems self-correct in real-time, achieving greater efficiency and reliability in wireless networks, including 5G and next-generation communication technologies.

Today's Problem and Its Solution

Problem:
Machine learning algorithms and communication networks face challenges in managing noisy, incomplete, or high-interference datasets. Autonomous vehicles, predictive analytics, and other AI systems often struggle with decision-making accuracy in unpredictable environments. Similarly, wireless communication systems, particularly 5G and beyond, encounter inefficiencies due to noise and signal interference. Current solutions involve increasing data redundancy or computational overhead, which is inefficient and limits scalability. These approaches fail to dynamically adapt to real-time disruptions, reducing reliability and performance in critical applications.

Solution:
By redefining noise as turbulence and modeling it using TDM, this invention introduces adaptive mechanisms for AI and communication systems. Machine learning algorithms simulate interdimensional energy flows to optimize pattern recognition and decision-making, stabilizing turbulent datasets. Communication systems adopt TDM-based strategies to self-correct in real-time, reducing reliance on redundancy and improving signal efficiency. These advancements lead to more accurate AI systems, resilient neural networks, and reliable, high-efficiency wireless communication infrastructure.

Detailed Description

1. **Conceptual Framework**

- **Turbulence in TDM:** Noise and interference are treated as turbulence within the twilight dimension, affecting the stability of energy flows. These dynamics are modeled and mitigated using algorithms inspired by TDM principles.
- **Stabilization Mechanism:** Neural networks and communication systems incorporate turbulence mitigation strategies, stabilizing energy flows and improving performance in noisy or incomplete datasets.

2. **System Components**
    - **TDM-Based Machine Learning Algorithms:** Algorithms simulate interdimensional energy flows, optimizing neural network performance by mitigating turbulence in noisy datasets. These models adapt dynamically to changing input conditions.
    - **Adaptive Communication Engine:** A hardware/software module that applies real-time turbulence modeling to enhance signal quality and reduce interference in wireless networks.
    - **Noise-to-Turbulence Translator:** A system module that converts conventional noise metrics into turbulence patterns, enabling targeted corrections and efficient data transmission.

3. **Applications**
    - **Autonomous Vehicles:** Machine learning models for navigation and decision-making achieve higher accuracy by stabilizing turbulent datasets, ensuring safe and efficient operation in unpredictable environments.
    - **Predictive Analytics:** AI systems enhance pattern recognition and adaptability in fields like finance, healthcare, and climate modeling by adopting TDM-inspired algorithms.
    - **Wireless Communication Networks:** TDM-based adaptive communication systems improve efficiency and reliability in 5G and beyond, transforming real-time data transmission in high-interference environments.

4. **Operational Workflow**
    - **Data Capture and Turbulence Mapping:** The system captures input data and maps turbulence patterns using TDM-based algorithms. These patterns are analyzed to identify sources of instability.
    - **Stabilization and Correction:** Machine learning algorithms and communication systems dynamically adjust to turbulence, stabilizing energy flows and ensuring accurate decision-making or efficient signal transmission.
    - **Continuous Adaptation:** The system continuously learns from turbulence patterns, refining its algorithms to enhance performance over time.

5. **Prototype Implementation**
    - **Machine Learning:** A neural network model trained on TDM-inspired turbulence simulations demonstrates improved accuracy in noisy environments, outperforming conventional algorithms.
    - **Wireless Communication:** A prototype communication system integrates TDM-based real-time correction, achieving enhanced signal quality and reduced latency in high-interference settings.

6. **Advantages**
    - **Enhanced AI Performance:** Machine learning models achieve higher accuracy and adaptability by mitigating turbulence in noisy datasets.

- **Real-Time Adaptation:** Communication systems dynamically self-correct, reducing data loss and improving reliability.
- **Resource Efficiency:** By minimizing reliance on redundancy, the system reduces computational and data overhead, enabling scalable and efficient solutions.

Claims

1. **TDM-Based Machine Learning Optimization System:**
   - A system that enhances machine learning algorithms by simulating interdimensional energy flows and mitigating turbulence in noisy or incomplete datasets. The system dynamically stabilizes neural network performance, improving pattern recognition, decision-making, and adaptability across various applications.
2. **Adaptive Communication Engine:**
   - A hardware/software module that applies TDM-based turbulence modeling to wireless communication networks, enabling real-time self-correction and signal stabilization. The module reduces reliance on redundancy, improving data transmission efficiency and reliability in high-interference environments.
3. **Noise-to-Turbulence Translation Framework:**
   - A framework that converts conventional noise metrics into turbulence patterns for targeted correction. This framework integrates with machine learning models and communication systems, optimizing performance by stabilizing interdimensional energy flows.

Provisional Patent Application: Twilight Dimension Model-Based Cryptography for Enhanced Data Security

Introduction

This invention presents a cryptographic system based on the Twilight Dimension Model (TDM), offering unparalleled data security by leveraging the unique dynamics of interdimensional turbulence. Traditional and quantum encryption techniques rely on computational complexity or quantum principles to secure communications, but they face vulnerabilities as computational power advances. TDM-based cryptography introduces a new paradigm by utilizing the unpredictable turbulence patterns of the twilight dimension to create secure, non-replicable encryption keys. This approach establishes a highly robust cryptographic framework for financial systems, government communications, and personal data protection, providing superior security even against quantum decryption technologies.

Today's Problem and Its Solution

Problem:
The increasing sophistication of computational power and quantum technologies threatens the security of existing cryptographic methods. Classical encryption techniques, such as RSA or AES, rely on complex algorithms, which are becoming increasingly vulnerable to quantum attacks. Quantum encryption, while more secure, still faces challenges such as scalability, implementation complexity, and susceptibility to physical-layer attacks. These limitations put critical data—such as financial transactions, government communications, and personal information—at risk of exposure, theft, or manipulation.

Solution:
TDM-based cryptography provides a transformative solution by leveraging the unique properties of interdimensional turbulence. Unlike traditional or quantum encryption, TDM-generated keys are based on non-reproducible turbulence patterns, ensuring unmatched security. These keys dynamically adapt to communication environments, creating a cryptographic system resistant to brute force, quantum decryption, and physical-layer attacks. This innovation establishes a scalable, efficient, and secure method for protecting sensitive data across various sectors.

Detailed Description

1. **Conceptual Framework**
    - **Twilight Dimension and Turbulence:** TDM models the twilight dimension as a realm of potential states activated into reality through interdimensional energy flows. Turbulence within this dimension introduces inherent unpredictability, which serves as the foundation for generating encryption keys.

- 2. **System Components**
    - **Dynamic Key Generation:** The cryptographic system generates unique, non-reproducible encryption keys by mapping turbulence patterns in real-time, ensuring that each key is specific to the moment and communication environment.
- 2. **System Components**
    - **Turbulence Detection Module:** Advanced sensors capture turbulence patterns within the twilight dimension, translating them into high-resolution datasets.
    - **Key Generation Engine:** Algorithms process turbulence data to extract unique features and encode them into cryptographic keys.
    - **Adaptive Encryption and Decryption Modules:** These modules integrate turbulence-based keys into encryption protocols, ensuring compatibility with existing communication systems while enhancing security.
- 3. **Operational Workflow**
    - **Initialization:** The system initializes by activating a specific set of potential states in the twilight dimension, generating interdimensional turbulence.
    - **Turbulence Mapping:** The turbulence detection module records real-time patterns, capturing their chaotic and non-linear characteristics.
    - **Key Generation:** The key generation engine processes the mapped turbulence to create secure encryption keys unique to the transaction or communication session.
    - **Encryption and Transmission:** Data is encrypted using the turbulence-based keys and transmitted securely.
    - **Decryption:** Authorized receivers use synchronized turbulence mapping to recreate the encryption key and decrypt the data.
- 4. **Applications**
    - **Financial Systems:** Protects high-value transactions by generating unique keys resistant to quantum and computational decryption.
    - **Government Communications:** Ensures secure data transmission in sensitive diplomatic and military communications.
    - **Personal Data Protection:** Safeguards personal information against emerging threats in an increasingly digital landscape.
- 5. **Advantages**
    - **Quantum Resistance:** Turbulence-based keys are inherently resistant to quantum decryption due to their unpredictability and non-reproducibility.
    - **Dynamic Security:** The system generates unique keys for each communication session, reducing vulnerability to interception or replication.
    - **Efficiency and Scalability:** The cryptographic framework minimizes computational overhead, enabling secure communication across diverse applications and infrastructures.

Claims

1. **TDM-Based Cryptographic Key Generation System:**
    - A system for generating encryption keys by mapping and processing turbulence patterns within the twilight dimension. The system includes a turbulence detection

module, a key generation engine, and adaptive encryption modules, providing unparalleled security for data transmission and storage.
2. **Dynamic Encryption Algorithm Using Turbulence Patterns:**
   - An algorithm that processes real-time turbulence data to create unique, non-reproducible encryption keys. The algorithm ensures resistance to brute force and quantum decryption techniques by leveraging the chaotic nature of interdimensional turbulence.
3. **Adaptive Cryptographic Framework:**
   - A cryptographic framework integrating TDM-based encryption keys with existing protocols. The framework includes synchronized turbulence mapping for real-time decryption, ensuring compatibility and enhanced security in applications such as financial systems, government communications, and personal data protection.

Provisional Patent Application: Enhancing Machine Learning with Turbulence Dynamics from the Twilight Dimension Model (TDM)

Introduction

This invention leverages the Twilight Dimension Model (TDM) to enhance machine learning systems by integrating the dynamics of turbulence during the "exploration" phase of learning. In TDM, turbulence describes the chaotic interaction of potential states before a single state is activated, aligning closely with how machine learning algorithms analyze and refine patterns from complex, noisy datasets. By modeling turbulence, this invention introduces a new framework for optimizing reinforcement learning, enabling systems to explore more diverse solutions and achieve superior accuracy, even when dealing with incomplete or noisy data.

The invention holds particular promise for applications such as financial market prediction, autonomous navigation, and healthcare analytics, where improved adaptability and robustness are critical.

Today's Problem and Its Solution

Problem:
Machine learning systems often struggle to manage noisy or incomplete datasets, leading to reduced accuracy in decision-making and predictions. Traditional models use brute-force or redundancy-based techniques to handle uncertainty, which are computationally expensive and inefficient. Reinforcement learning, while adaptive, typically explores a limited solution space due to its trial-and-error approach, restricting its effectiveness in dynamic, real-world scenarios.

Solution:
By incorporating TDM's turbulence dynamics, machine learning algorithms can simulate the chaotic interactions of potential states, mirroring the complexity of real-world data. This approach enhances the "exploration" phase of learning, enabling systems to identify and evaluate diverse solutions before selecting an optimal state. As a result, AI systems become more resilient to noise, better equipped to handle incomplete datasets, and capable of generating more accurate predictions.

Detailed Description

1. **Conceptual Framework**
    - **Turbulence in TDM:** Turbulence represents the chaotic interaction of potential states within the twilight dimension. These interactions introduce randomness and

complexity, which are resolved into a single, activated state in the reality dimension.
   - **Machine Learning Analog:** In machine learning, turbulence aligns with the exploration of various hypotheses or solutions before settling on an optimal model. By simulating turbulence, algorithms can navigate and refine patterns in complex datasets more effectively.
2. **System Components**
   - **Turbulence Modeling Engine:** Simulates TDM-inspired turbulence dynamics to introduce controlled randomness during the exploration phase of machine learning.
   - **Adaptive Optimization Layer:** Analyzes turbulence patterns to refine models dynamically, balancing exploration and exploitation.
   - **Noise Resilience Module:** Uses turbulence-based insights to improve system performance on incomplete or noisy datasets.
3. **Operational Workflow**
   - **Initialization:** The system initializes by mapping the data landscape and simulating turbulence dynamics using TDM principles.
   - **Exploration Phase:** Turbulence modeling generates diverse hypotheses by introducing controlled randomness into the learning process.
   - **Evaluation and Activation:** The system evaluates turbulence-induced hypotheses, activating the optimal state for the given data.
   - **Adaptation:** The optimization layer refines the model based on feedback, iterating the process for improved performance.
4. **Applications**
   - **Financial Market Analysis:** Simulating chaotic market interactions to improve trend prediction and investment strategies.
   - **Autonomous Navigation:** Enhancing decision-making in dynamic environments by modeling turbulence to anticipate and adapt to unforeseen variables.
   - **Healthcare Analytics:** Processing incomplete patient data to identify potential diagnoses or treatment plans with greater accuracy.
5. **Advantages**
   - **Enhanced Exploration:** Turbulence modeling allows systems to explore a wider range of potential solutions, improving adaptability.
   - **Improved Noise Resilience:** Algorithms perform better on noisy or incomplete datasets by leveraging turbulence-based insights.
   - **Efficiency and Scalability:** The framework reduces computational overhead by focusing on structured exploration rather than brute force.

Claims

1. **Turbulence Modeling Engine for Machine Learning Systems:**
   - A turbulence modeling engine that integrates TDM dynamics into machine learning systems, enabling enhanced exploration of potential solutions by simulating chaotic interactions akin to turbulence in the twilight dimension. The

engine generates diverse hypotheses during the exploration phase, improving system adaptability and accuracy.
2. **Adaptive Optimization Layer with Turbulence Dynamics:**
   - An adaptive optimization layer that analyzes turbulence patterns to refine machine learning models dynamically. This layer balances exploration and exploitation, ensuring optimal decision-making in environments characterized by noise or incomplete datasets.
3. **Noise Resilience Module Utilizing TDM Principles:**
   - A noise resilience module that applies turbulence-based insights to improve the robustness of machine learning systems. By simulating and resolving chaotic interactions, the module enhances the system's ability to handle noisy or incomplete data, achieving superior accuracy and performance.

Provisional Patent Application: Resistance-Based Bias Mitigation in AI Systems Inspired by the Twilight Dimension Model (TDM)

Introduction

This invention introduces a novel method for mitigating bias in AI systems using the Twilight Dimension Model (TDM). Bias in AI arises when algorithms favor certain outcomes disproportionately due to imbalanced training data or flawed decision-making processes. TDM introduces the concept of resistance, a mechanism that impedes state activation, which can be adapted to counteract biased tendencies in AI. By simulating resistance as a counterforce, AI systems can ensure a more balanced exploration of solutions and reduce systemic biases. Applications include AI-driven hiring processes, predictive analytics, and other domains where fairness and objectivity are critical.

Today's Problem and Its Solution

Problem:
Bias in AI systems is a pervasive issue, particularly in applications like hiring, loan approvals, and criminal justice. Algorithms trained on imbalanced datasets often perpetuate systemic inequalities by favoring the dominant patterns in the data, leading to unfair outcomes. Additionally, AI systems are prone to overfitting, where they become overly focused on specific patterns, reducing their ability to generalize across diverse datasets. Current bias mitigation techniques often require manual intervention or result in compromised model performance, which limits their scalability and efficacy.

Solution:
By incorporating TDM-inspired resistance into AI models, this invention addresses both bias and overfitting. Resistance serves as a counterforce to overly dominant tendencies in data or decision-making, promoting balanced exploration and ensuring fairness. For instance, in hiring algorithms, resistance can amplify underrepresented candidate profiles, reducing systemic biases. In broader applications, resistance introduces penalties for overly simplistic solutions, preventing overfitting and enhancing the model's ability to generalize. This approach combines fairness with improved robustness and adaptability, offering a scalable solution to bias mitigation in AI.

Detailed Description

1. **Conceptual Framework**
    - **Resistance in TDM:** In TDM, resistance acts as a counterforce that impedes state activation, ensuring equilibrium in energy flows. This concept is adapted to AI by simulating resistance to counteract biased tendencies and overly dominant patterns in the data.

- **Bias Mitigation Mechanism:** Resistance is integrated into AI models as a dynamic parameter that penalizes decisions disproportionately influenced by dominant biases, encouraging a more balanced exploration of potential solutions.

2. **System Components**
    - **Resistance Modeling Engine:** Simulates resistance dynamics based on TDM principles, generating counterforces to mitigate bias in AI models.
    - **Bias Detection Module:** Identifies biased tendencies in data or decision-making processes, providing real-time feedback to the resistance modeling engine.
    - **Adaptive Penalty Layer:** Applies resistance-based penalties during model training, ensuring that the AI system avoids overfitting and promotes fairness.

3. **Operational Workflow**
    - **Data Analysis:** The bias detection module analyzes training data to identify imbalances and potential sources of bias.
    - **Resistance Simulation:** The resistance modeling engine introduces counterforces based on identified biases, dynamically adjusting during the learning process.
    - **Model Training:** The adaptive penalty layer integrates resistance dynamics into the optimization function, penalizing biased or overly simplistic solutions.
    - **Validation and Refinement:** The system evaluates model performance on fairness and generalization metrics, refining resistance parameters as needed.

4. **Applications**
    - **Hiring Processes:** AI systems integrate resistance to amplify underrepresented candidate profiles, ensuring equitable consideration.
    - **Predictive Analytics:** Resistance mitigates biases in data-driven predictions, improving fairness in applications like healthcare and finance.
    - **General AI Systems:** Resistance introduces penalties for overfitting, enhancing robustness and adaptability across diverse datasets.

5. **Advantages**
    - **Enhanced Fairness:** By counteracting biased tendencies, the system promotes equitable decision-making across various applications.
    - **Improved Generalization:** Resistance-based penalties prevent overfitting, enabling models to perform well on diverse and unseen datasets.
    - **Scalability:** The system is adaptable to different AI architectures and domains, offering a versatile solution to bias mitigation.

Claims

1. **Resistance-Based Bias Mitigation Engine:**
    - A system for mitigating bias in AI models by simulating resistance dynamics inspired by the Twilight Dimension Model. The engine generates counterforces to biased tendencies, ensuring balanced exploration of solutions and promoting fairness in decision-making processes.
2. **Adaptive Penalty Layer with Resistance Dynamics:**
    - An adaptive penalty layer that integrates resistance-based penalties into AI optimization functions. This layer dynamically adjusts penalties during training to

counteract bias and prevent overfitting, improving model fairness and generalization.
3. **Bias Detection and Resistance Integration Framework:**
    - A framework that combines bias detection with resistance modeling to dynamically mitigate bias in AI systems. The framework includes real-time feedback loops that refine resistance parameters, ensuring scalable and efficient bias mitigation across diverse applications.

Provisional Patent Application: Adaptive Artificial Intelligence Systems Inspired by Twilight Dimension Model (TDM)

Introduction

This invention leverages the Twilight Dimension Model (TDM) to create adaptive artificial intelligence (AI) systems capable of dynamically responding to new environments, challenges, and user preferences. Adaptability is a cornerstone of intelligent systems, and TDM's emphasis on dynamic energy flows provides a blueprint for enhancing the flexibility and robustness of AI. By mimicking interdimensional energy flows, AI systems can recalibrate their activation dynamics in real time, enabling seamless adjustments to unpredictable conditions. Applications include robotics for disaster response, personalized AI solutions, and systems designed for dynamic environments like healthcare or autonomous vehicles.

Today's Problem and Its Solution

Problem:
Existing AI systems often lack the adaptability required to handle novel or unpredictable environments effectively. Robotics, for instance, may fail in disaster scenarios due to rigid programming that cannot accommodate dynamic conditions like collapsing structures or shifting terrain. Similarly, AI-driven personalization systems often rely on static algorithms, limiting their ability to continuously evolve in response to user preferences. This lack of flexibility reduces the efficiency, reliability, and user satisfaction associated with AI applications.

Solution:
The proposed adaptive AI system draws on TDM's interdimensional energy flow dynamics to enable real-time recalibration of state activation processes. This approach equips AI systems with the ability to navigate unpredictable environments, recalibrate behaviors, and personalize responses dynamically. In robotics, for example, TDM-inspired adaptability allows systems to adjust to complex disaster conditions, enhancing their operational efficiency. For personalized AI, the framework enables continuous learning and adaptation to individual user needs, fostering improved user experiences and outcomes.

Detailed Description

1. **Conceptual Framework**
    - **TDM Dynamics and Adaptation:** TDM describes a system where potential states interact through energy flows, with dynamic adjustments to turbulence and resistance leading to state activation. AI systems mimic this process by modeling activation dynamics that adjust to changing inputs and environments.

- **Adaptive State Activation:** By incorporating TDM principles, AI systems recalibrate their internal state activation processes in response to external stimuli, enabling adaptability to new conditions or preferences.
2. **System Components**
    - **State Activation Module:** Simulates TDM-like energy flows to adjust internal AI states dynamically based on real-time inputs.
    - **Environmental Feedback Loop:** Continuously monitors external conditions, providing data for recalibration.
    - **Personalization Engine:** Analyzes user interactions to refine responses, ensuring individualized adaptability over time.
3. **Operational Workflow**
    - **Initialization:** The system initializes by mapping the current environment or user context, establishing baseline state activation processes.
    - **Dynamic Adaptation:** The state activation module adjusts activation pathways dynamically in response to external inputs or environmental changes, such as obstacles or user feedback.
    - **Feedback Integration:** The environmental feedback loop and personalization engine continuously refine the system's responses, ensuring optimal performance in dynamic or unpredictable scenarios.
4. **Applications**
    - **Disaster Response Robotics:** Robots use TDM-inspired adaptability to recalibrate behaviors in response to shifting terrain or structural collapse, enabling efficient navigation and survivor location.
    - **Healthcare Personalization:** AI systems dynamically adapt treatment recommendations based on patient-specific data and changing health conditions.
    - **Autonomous Navigation:** Vehicles adjust their decision-making processes in real time, accommodating dynamic road conditions and obstacles.
5. **Advantages**
    - **Enhanced Flexibility:** The system dynamically adapts to new conditions, improving reliability in unpredictable environments.
    - **Continuous Personalization:** By continuously learning and recalibrating, the system ensures individualized responses to user preferences or needs.
    - **Robust Performance:** TDM-inspired dynamics enhance the system's ability to operate effectively across diverse applications and scenarios.

Claims

1. **Adaptive State Activation System:**
    - A system for dynamically adjusting internal activation processes in AI models based on the principles of interdimensional energy flows described in the Twilight Dimension Model. The system includes a state activation module that recalibrates pathways in response to environmental or user-specific inputs.
2. **Environmental Feedback Loop for Adaptive AI:**

- A feedback loop that continuously monitors external conditions and provides real-time data for recalibration of AI state activation processes. This loop ensures robust adaptability in dynamic environments, including disaster scenarios and autonomous navigation.
3. **Personalization Engine with Dynamic Adaptation:**
    - A personalization engine that refines AI system responses by analyzing user interactions and preferences. By integrating TDM-inspired adaptability, the engine ensures continuous learning and customization to individual user needs.

Provisional Patent Application: Turbulence-Based AI for Enhanced Search and Recommendation Systems

Introduction

This invention introduces an AI framework inspired by the Twilight Dimension Model (TDM) for improving the diversity and relevance of results in search engines and recommendation systems. By modeling turbulence during the exploration phase, the system identifies connections and solutions that might otherwise remain hidden in conventional algorithms. This dynamic approach expands result diversity, reduces biases, and enhances user satisfaction, transforming how digital platforms curate and present content.

Today's Problem and Its Solution

Problem:
Search engines and recommendation systems often rely on static algorithms that prioritize popular or easily identifiable patterns in data. This approach leads to homogeneity in results, limiting diversity and overlooking niche or meaningful options. Current systems also suffer from biases, as they amplify dominant trends without accounting for latent variables in the data.

Solution:
The proposed system integrates TDM-inspired turbulence modeling to explore complex and dynamic relationships in datasets. By simulating the chaotic interaction of potential states, the system identifies diverse patterns and offers more meaningful results. This framework ensures broader exploration and balances relevance with diversity, leading to improved user engagement and satisfaction.

Detailed Description

1. **Conceptual Framework**
   - **TDM and Turbulence Dynamics:** TDM-inspired turbulence represents the chaotic interplay of data states. This dynamic is adapted to AI algorithms, allowing them to navigate non-linear relationships in datasets.
   - **Explorative Diversity:** Turbulence introduces controlled randomness during the exploration phase, enabling algorithms to identify diverse and less obvious connections.
2. **System Components**
   - **Turbulence Modeling Engine:** Generates turbulence simulations to introduce variability during the data exploration phase.
   - **Dynamic Ranking Layer:** Adjusts search or recommendation rankings based on turbulence-derived insights to prioritize diverse results.

- **Bias Mitigation Module:** Counteracts dominant trends by integrating turbulence-based penalties, ensuring balanced exploration.
3. **Applications**
    - **Search Engines:** Provides users with a mix of mainstream and niche results by dynamically exploring non-linear relationships in search queries.
    - **Recommendation Systems:** Enhances content discovery in platforms like e-commerce, video streaming, and social media by broadening the spectrum of recommendations.

Claims

1. **Turbulence-Based Exploration Engine:**
    - A system for integrating turbulence modeling into AI algorithms to enhance the diversity of search and recommendation results, reducing homogeneity and amplifying meaningful discoveries.
2. **Dynamic Ranking and Bias Mitigation Layer:**
    - A layer that applies turbulence-based penalties to dominant trends in data, ensuring balanced exploration during the recommendation process.
3. **Search and Recommendation Optimization Framework:**
    - A framework combining turbulence modeling and dynamic ranking mechanisms to improve user engagement through diversified and meaningful search results.

Provisional Patent Application: AI Systems for Personalized Healthcare Using Resistance Modeling

Introduction

This invention applies TDM principles to AI-driven healthcare systems by incorporating resistance modeling to tailor treatment plans. Resistance, as defined in TDM, represents factors impeding state activation, which can analogously reflect patient-specific challenges. The system refines treatment recommendations dynamically, addressing individual health profiles and evolving conditions.

Today's Problem and Its Solution

Problem:
Conventional AI-driven healthcare systems often rely on generalized models that fail to consider patient-specific nuances. This lack of personalization can result in ineffective or suboptimal treatment recommendations. Additionally, static models struggle to adapt to evolving health conditions, limiting their utility in real-world scenarios.

Solution:
The proposed system introduces TDM-inspired resistance modeling to account for patient-specific variables. By dynamically adapting to health data, the system generates tailored treatment plans that evolve with the patient's condition. This ensures precise, personalized, and effective healthcare recommendations.

Detailed Description

1. **Conceptual Framework**
   - **Resistance Modeling in TDM:** Resistance factors represent obstacles to state activation. In healthcare, these correspond to patient-specific challenges, such as comorbidities or genetic predispositions.
   - **Dynamic Adaptation:** The system adjusts treatment plans in real time based on resistance modeling, ensuring continuous personalization.
2. **System Components**
   - **Patient Resistance Profiling Module:** Maps resistance factors unique to each patient, such as medical history, genetic data, and environmental influences.
   - **Adaptive Treatment Generator:** Recommends treatments dynamically, adjusting to resistance profiles and evolving health conditions.
   - **Feedback Integration Layer:** Refines recommendations based on treatment outcomes and patient feedback.
3. **Applications**

- **Chronic Disease Management:** Personalizes long-term care strategies for conditions like diabetes or cardiovascular diseases.
- **Precision Medicine:** Enhances the effectiveness of treatments by integrating genetic and environmental data into resistance modeling.

Claims

1. **Resistance Modeling Engine for Healthcare AI:**
   - A system for mapping patient-specific resistance factors to dynamically tailor treatment recommendations, improving personalization and efficacy.
2. **Adaptive Treatment Generator with Resistance Dynamics:**
   - A module that integrates resistance profiles into treatment algorithms, ensuring real-time adjustments based on patient data.
3. **Feedback-Driven Personalization Framework:**
   - A framework combining resistance modeling and feedback integration to refine healthcare recommendations dynamically.

www.ingramcontent.com/pod-product-compliance
Lightning Source LLC
Chambersburg PA
CBHW051151220526
45473CB00003B/729